AF361861

Nanoscale Thermodynamics

Nanoscale Thermodynamics

Editor

Signe Kjelstrup

MDPI • Basel • Beijing • Wuhan • Barcelona • Belgrade • Manchester • Tokyo • Cluj • Tianjin

Editor
Signe Kjelstrup
Norwegian University of Science and Technology
Norway

Editorial Office
MDPI
St. Alban-Anlage 66
4052 Basel, Switzerland

This is a reprint of articles from the Special Issue published online in the open access journal *Nanomaterials* (ISSN 2079-4991) (available at: https://www.mdpi.com/journal/nanomaterials/special_issues/nanoscale_thermodynamics).

For citation purposes, cite each article independently as indicated on the article page online and as indicated below:

LastName, A.A.; LastName, B.B.; LastName, C.C. Article Title. *Journal Name* **Year**, *Volume Number*, Page Range.

ISBN 978-3-0365-1168-9 (Hbk)
ISBN 978-3-0365-1169-6 (PDF)

Contents

About the Editor

Signe Kjelstrup, prof. dr.techn. et dr.ing., is a principal investigator at the PoreLab Center of Excellence, Norwegian University of Science and Technology. She has written monographs on non-equilibrium thermodynamics for heterogeneous systems and non-equilibrium thermodynamics for engineers, and now also nanothermodynamics. She has written about 350 journal articles and is a member of The Norwegian Academy of Sciences, Norwegian Academy of Technical Sciences, and Academia Europa.]

 *nanomaterials*

MDPI

Editorial

Special Issue on Nanoscale Thermodynamics

Signe Kjelstrup

PoreLab, Department of Chemistry, Faculty of Natural Sciences, Norwegian University of Science and Technology, 7491 Trondheim, Norway; signe.kjelstrup@ntnu.no

check for **updates**

Citation: Kjelstrup, S. Special Issue on Nanoscale Thermodynamics. *Nanomaterials* **2021**, *11*, 584. https://doi.org/10.3390/nano11030584

Received: 20 February 2021
Accepted: 23 February 2021
Published: 26 February 2021

This Special Issue concerns recent developments of a theory for energy conversion on the nanoscale, namely nanothermodynamics. The theory applies to porous media, small surfaces, clusters or fluids under confinement. There are a large number of unsolved issues in these contexts and present efforts only paint part of the broader picture. We may still ask questions on how far down in scale we can really use the Gibbs equation. Which theory can replace the Gibbs equation beyond the thermodynamic limit?

It is well known that confinement can change the equation of the state of a fluid, but how does confinement change the equilibrium conditions? How do we formulate equilibrium conditions on the nanoscale, and what are the independent variables? To deal with equilibrium alone seems a formidable task, let alone how to extend the descriptions to systems away from equilibrium.

This Special Issue explores in more detail some roads that were opened by Hill when he launched his thermodynamics for small systems in 1963. His method has, however, not gained much attention since it was published. We now consider this an underused opportunity. The theoretical developments in nanothechnology need to follow the experimental progress, and that is rapid. It is our ambition, therefore, to aspire to an increased effort that can further develop suitable theoretical tools and methods in nanoscience. All ten contributions to this Special Issue can be seen as efforts to support, enhance and validate such theoretical developments.

The first two papers [1,2] demonstrate the use of Hill's nanothermodynamics in new settings. The Small System Method for the determination of thermodynamic factors has already been successfully applied in many contexts. The method exploits the small system's scaling properties or size dependencies. In this Special Issue, Dawass et al. [1] show how the analysis of Kirkwood Buff integrals can be made more accurate. They conclude that this is possible, by applying three methods to compute these integrals in the thermodynamic limit. Radial distribution functions (RDFs) of finite systems are used. Tripathy et al. [2] extend the Small System Method further to also characterize the hydration shell compressibility of a generally hydrophobic polymer in water. They show how this finding may be generalized to study hydrophobic interactions.

The next three papers [3–5] concern the pressure of confined fluids and their description in equilibrium. Rauter et al. show [3] that the integral pressure is constant across phase boundaries and that this finding is equivalent to assuming validity of Young's and Young–Laplace's law. In agreement with this, Máté Erdős et al. [4] document the interrelation of the differential and the integral pressure. Galteland et al. [5] show how the disjoining pressure can be understood using Hill's theory, and present Maxwell relations for small systems.

The issue contains three extensions of Hill's theory, see papers [6–8]. Hill stated that small systems do not obey Legendre transforms, a clear disadvantage. Beering et al. [6] have been able to show for the first time in their article "*A Legendre–Fenchel Transform for Molecular Stretching Energies*" that an alternative for small systems lies in the Legendre–Fenchel transform. This is an extension of Hill's theory that may prove useful in practice. Strøm et al. [7] have been able to extend on Hill's examples, by considering adsorbed films on very small clusters. Their article "*When Thermodynamic Properties of Adsorbent Films*

Depends on Size" offers a new way to deal with adsorption on, say, atmospheric particles. In yet another article, Strøm et al. [8] were able to compute and illustrate the equation of state of an ideal gas when it becomes confined. Large discrepancies from normal ideal gas behavior are found. The equation of state can be computed exactly for an ideal gas using statistical mechanics and is illustrated by molecular dynamics simulations.

The last two papers [9,10] are special. The only experimental paper in this issue is provided by Men'shikov et al. [9]. The authors show how sensitive the structure is to environmental conditions, and how the results for adsorption enthalpy vary with temperature, carbon porosity and surface area. Therefore, there are clear indications of multilayer formations on some of their activated carbons, and abnormal effects, which may benefit from other theories. So far, the treatment of these data has followed the classical scheme. Is there an alternative route, simpler than that of Hill, to nanothermodynamics? Rodrigo de Miguel and J. Miguel Rubi [10] propose this in their study on *"Statistical Mechanics at Strong Coupling: A Bridge between Landsberg's Energy Levels and Hill's Nanothermodynamics"*. They review and show the connection between three theories, including Hill's, proposed for the thermodynamic treatment of systems that do not obey the additivity ansatz of classical thermodynamics.

Funding: S.K. acknowledges The Research Council of Norway under the Centre of Excellence funding scheme, project PoreLab, project no 262644.

Institutional Review Board Statement: Not applicable.

Informed Consent Statement: Not applicable.

Acknowledgments: The guest editor thanks all the authors for submitting their work to the Special Issue and the editorial office for its successful transformation into a physical issue. A special thank you to all reviewers participating in the peer-review process of the submitted manuscripts, for enhancing their quality and impact.

Conflicts of Interest: Review proceedings of the contributions coauthored by S. Kjelstrup were handled by reviewers and academic editors.

References

1. Dawass, N.; Krüger, P.; Schnell, S.K.; Moultos, O.A.; Economou, I.G.; Vlugt, T.J.H.; Simon, J.-M. Kirkwood-Buff Integrals Using Molecular Simulation: Estimation of Surface Effects. *Nanomaterials* **2020**, *10*, 771. [CrossRef] [PubMed]
2. Tripathy, M.; Bharadwaj, S.; B., S.J.; van der Vegt, N.F.A. Characterizing Polymer Hydration Shell Compressibilities with the Small-System Method. *Nanomaterials* **2020**, *10*, 1460. [CrossRef] [PubMed]
3. Erdős, M.; Galteland, O.; Bedeaux, D.; Kjelstrup, S.; Moultos, O.A.; Vlugt, T.J.H. Gibbs Ensemble Monte Carlo Simulation of Fluids in Confinement: Relation between the Differential and Integral Pressures. *Nanomaterials* **2020**, *10*, 293. [CrossRef]
4. Rauter, M.T.; Galteland, O.; Erdős, M.; Moultos, O.A.; Vlugt, T.J.H.; Schnell, S.K.; Bedeaux, D.; Kjelstrup, S. Two-Phase Equilibrium Conditions in Nanopores. *Nanomaterials* **2020**, *10*, 608. [CrossRef] [PubMed]
5. Galteland, O.; Bedeaux, D.; Kjelstrup, S. Nanothermodynamic Description and Molecular Simulation of a Single-Phase Fluid in a Slit Pore. *Nanomaterials* **2021**, *11*, 165. [CrossRef] [PubMed]
6. Bering, E.; Bedeaux, D.; Kjelstrup, S.; de Wijn, A.S.; Latella, I.; Rubi, J.M. A Legendre–Fenchel Transform for Molecular Stretching Energies. *Nanomaterials* **2020**, *10*, 2355. [CrossRef] [PubMed]
7. Strøm, B.A.; He, J.; Bedeaux, D.; Kjelstrup, S. When Thermodynamic Properties of Adsorbed Films Depend on Size: Fundamental Theory and Case Study. *Nanomaterials* **2020**, *10*, 1691. [CrossRef]
8. Strøm, B.A.; Bedeaux, D.; Schnell, S.K. Adsorption of an Ideal Gas on a Small Spherical Adsorbent. *Nanomaterials* **2021**, *11*, 431. [CrossRef]
9. Men'shchikov, I.; Shkolin, A.; Khozina, E.; Fomkin, A. Thermodynamics of Adsorbed Methane Storage Systems Based on Peat-Derived Activated Carbons. *Nanomaterials* **2020**, *10*, 1379. [CrossRef]
10. De Miguel, R.; Rubí, J.M. Statistical Mechanics at Strong Coupling: A Bridge between Landsberg's Energy Levels and Hill's Nanothermodynamics. *Nanomaterials* **2020**, *10*, 2471. [CrossRef] [PubMed]

 *nanomaterials* 

Article

Kirkwood-Buff Integrals Using Molecular Simulation: Estimation of Surface Effects

Noura Dawass [1], Peter Krüger [2], Sondre K. Schnell [3], Othonas A. Moultos [1], Ioannis G. Economou [4,5], Thijs J. H. Vlugt [1] and Jean-Marc Simon [6,*

1   Engineering Thermodynamics, Process & Energy Department, Faculty of Mechanical, Maritime and Materials Engineering, Delft University of Technology, Leeghwaterstraat 39, 2628CB Delft, The Netherlands; t.j.h.vlugt@tudelft.nl (T.J.H.V.)

2   Graduate School of Engineering and Molecular Chirality Research Center, Chiba University, Chiba 263-8522, Japan

3   Department of Materials Science and Engineering, NTNU, N-7491 Trondheim, Norway

4   National Center for Scientific Research Demokritos, Institute of Nanoscience and Nanotechnology, Molecular Thermodynamics and Modelling of Materials Laboratory, GR 153 10 Aghia Paraskevi Attikis, Greece

5   Chemical Engineering Program, Texas A&M University at Qatar, Education City, Doha PO Box 23874, Qatar

6   ICB, UMR 6303 CNRS-Université de Bourgogne, F-21078 Dijon, France

*   Correspondence: jmsimon@u-bourgogne.fr

Received: 13 March 2020; Accepted: 13 April 2020; Published: 16 April 2020

Abstract: Kirkwood-Buff (KB) integrals provide a connection between microscopic properties and thermodynamic properties of multicomponent fluids. The estimation of KB integrals using molecular simulations of finite systems requires accounting for finite size effects. In the small system method, properties of finite subvolumes with different sizes embedded in a larger volume can be used to extrapolate to macroscopic thermodynamic properties. KB integrals computed from small subvolumes scale with the inverse size of the system. This scaling was used to find KB integrals in the thermodynamic limit. To reduce numerical inaccuracies that arise from this extrapolation, alternative approaches were considered in this work. Three methods for computing KB integrals in the thermodynamic limit from information of radial distribution functions (RDFs) of finite systems were compared. These methods allowed for the computation of surface effects. KB integrals and surface terms in the thermodynamic limit were computed for Lennard-Jones (LJ) and Weeks-Chandler-Andersen (WCA) fluids. It was found that all three methods converge to the same value. The main differentiating factor was the speed of convergence with system size L. The method that required the smallest size was the one which exploited the scaling of the finite volume KB integral multiplied by L. The relationship between KB integrals and surface effects was studied for a range of densities.

Keywords: nanothermodynamics; Kirkwood-Buff integrals; surface effects; molecular dynamics

---

1. Introduction

Using knowledge of the molecular structure of liquids to predict their macroscopic behavior is important for several applications [1–5]. One of the most rigorous solution theories is the Kirkwood–Buff (KB) theory, where a sound connection between macroscopic and microscopic properties for isotropic multicomponent fluids is established [6,7]. Kirkwood and Buff derived a relation between several thermodynamic properties and integrals of radial distribution functions (RDFs) over infinite and open volumes $G_{\alpha\beta}^{\infty}$ in the grand-canonical ensemble [6]:

$$G_{\alpha\beta}^{\infty} = \int_0^{\infty} \left[g_{\alpha\beta}^{\infty}(r) - 1 \right] 4\pi r^2 \mathrm{d}r \tag{1}$$

where r is the particle distance, and $g_{\alpha\beta}^{\infty}(r)$ is the RDF, of the infinitely large system, for species α and β. KB integrals can also be expressed in terms of density fluctuations in open systems [6–9]. While KB integrals were derived for open and infinite systems, many studies use molecular simulation to estimate KB integrals, where only finite systems can be studied. In Reference [10], a review of the methods available in literature for computing KB integrals from molecular simulations is presented.

To accurately estimate $G_{\alpha\beta}^{\infty}$, it is possible to use KB integrals of finite and open subvolumes V embedded in larger reservoirs. In this way, the grand-canonical ensemble, in which KB integrals in the thermodynamic limit were derived, is mimicked. This approach is referred to as the small system method (SSM) [5,11,12]. According to the SSM, properties of small subvolumes, that can be of the order of a few molecular diameters, are treated in terms of thermodynamics of small systems rather than classical thermodynamics. According to Hill's thermodynamics of small systems, properties of open embedded subvolumes scale with the inverse size of the subvolumes [13,14]. This also applies to KB integrals of finite subvolumes, $G_{\alpha\beta}^{V}$ [10,12]. For a specific system, $G_{\alpha\beta}^{V}$ computed with subvolumes of different sizes, scales linearly with the inverse size of the subvolume [10,12,15,16]. For spherical subvolumes V inside a simulation box, we have

$$G_{\alpha\beta}^{V}(L) = G_{\alpha\beta}^{\infty} + \frac{F^{\infty}}{L} \tag{2}$$

where $L = 6V/A$ is the characteristic length of the subvolume V with surface area A. For a sphere, L is the diameter ($L = 2R$). In Equation (2), F^{∞} is related to surface effects of the subvolume. Using Hill's formulation of small-system thermodynamics [14], it was shown that properties of small systems can be written in terms of volume and surface contributions [17]. In Reference [17], Hill's thermodynamics were applied to several properties, including pressure. From the volume contribution of pressure, the homogeneous pressure is obtained, while the Gibbs surface relation was obtained from the surface contribution [17]. This last contribution is proportional to the surface tension. In the case of KB integrals, the surface term, or contribution, F^{∞}, can also be defined from Gibbs surface equation [17]. From a microscopic point of view, it originates from interactions between molecules inside the subvolume and molecules across the boundary of the subvolume [12,15]. These surface effects vanish in the thermodynamic limit, but for systems used in molecular dynamics (MD) simulations these effects cannot be neglected [18]. As a result, the quantitative and qualitative study of surface contributions is essential for estimating $G_{\alpha\beta}^{\infty}$ from integrals of finite subvolumes $G_{\alpha\beta}^{V}$.

KB integrals of finite and open subvolumes $G_{\alpha\beta}^{V}$ are defined in terms of fluctuations in the number of particles, which relate to double integrals of RDFs over the subvolume V [12],

$$G_{\alpha\beta}^{V} \equiv \frac{1}{V} \int_V \int_V \left[g_{\alpha\beta}(r_{12}) - 1 \right] \mathrm{d}\mathbf{r}_1 \mathrm{d}\mathbf{r}_2 \equiv V \frac{\langle N_\alpha N_\beta \rangle - \langle N_\alpha \rangle \langle N_\beta \rangle}{\langle N_\alpha \rangle \langle N_\beta \rangle} - \frac{V \delta_{\alpha\beta}}{\langle N_\beta \rangle} \tag{3}$$

where N_α and N_β are the number of molecules of type α and β, in volume V. The brackets $\langle \cdots \rangle$ denote an ensemble average in an open system. Equation (3) is applicable to isotropic molecular fluids where the orientations of molecules are already integrated out. While it is possible to compute KB integrals $G_{\alpha\beta}^{V}$ from fluctuations in the number of particles (i.e., the right hand side of Equation (3)), it is more practical to use RDFs. RDFs are readily computed by most molecular simulation software packages. The double integrals in the left hand side of Equation (3) can be transformed to a single integral using a weight function $w(r, L)$ [12],

$$G^V_{\alpha\beta} = \int_0^L \left[g_{\alpha\beta}(r) - 1 \right] w(r, L) \mathrm{d}r \tag{4}$$

The function $w(r, L)$ depends on the geometry of V. For spherical and cubic subvolumes, theoretically derived functions are available in References [12,16], respectively. It is possible to numerically obtain the function $w(r, L)$ for an arbitrary shape as shown in Reference [19]. In this work, spherical subvolumes will be used, for which,

$$w(r, L) = 4\pi r^2 \left(1 - \frac{3x}{2} + \frac{x^3}{2} \right) \tag{5}$$

where x is the dimensionless distance r/L [12,15,16]. The scaling of finite integrals G^V with the size of the subvolumes L is used to compute KB integrals in the thermodynamic limit G^∞ (for convenience, indicies α and β will be dropped from this point onwards). Specifically, G^∞ is computed from extrapolating the linear part of the scaling of G^V with $1/L$ to the limit $1/L \rightarrow 0$ [12,15,17]. A disadvantage of this approach is that a linear regime is not always easily identified [15].

To avoid extrapolating G^V, Krüger and Vlugt [16] proposed a direct estimation of KB integrals in the thermodynamic limit:

$$G^\infty \approx G_k(L) = \int_0^L \left[g(r) - 1 \right] u_k(r) \mathrm{d}r \tag{6}$$

The accuracy of the estimation depends on the function $u_k(r)$ [20]. Krüger and Vlugt [16] considered three different estimations and found that integrals computed using the function $u_2(r)$ provided the best estimation of G^∞,

$$u_2(r) = 4\pi r^2 \left(1 - \frac{23}{8}x^3 + \frac{3}{4}x^4 + \frac{9}{8}x^5 \right) \tag{7}$$

KB integrals computed using Equations (6) and (7) will be denoted by G_2. To derive the expression for G_2, the starting point was the scaling of KB integrals with $1/L$. First, an explicit estimation of F^∞ in Equation (2) was derived. In the work of Krüger and Vlugt [16], F^∞ has the following form

$$F^\infty = \int_0^\infty [g(r) - 1] 4\pi r^2 \left(-\frac{3}{2}r \right) \mathrm{d}r \tag{8}$$

It is important to note that the structure of Equation (8) is similar to KB integrals in the thermodynamic limit (Equation (1)). So, analogous to Equation (2), F^V can be defined as,

$$F^V(L) = F^\infty + \frac{C}{L} \tag{9}$$

where C is a constant. For finite systems, F^V can be computed using

$$F^V \approx \int_0^L [g(r) - 1] \left(-\frac{3}{2}r \right) w(r, L) \mathrm{d}r \tag{10}$$

where the function $w(r, L)$ is given in Equation (5). The similarity between the expression for KB integrals (Equation (4)) and surface term (Equation (8)) in the thermodynamic limit allows for deriving an estimation for surface effects as in Equation (6). Using Equation (6), and Equation (8) an explicit expression for surface effects in the thermodynamic limit, denoted here by F_2^∞, is obtained from

$$F_2^\infty \approx \int_0^L [g(r) - 1] \left(-\frac{3}{2}r \right) u_2(r) \mathrm{d}r \tag{11}$$

with $u_2(r)$ in Equation (7).

An alternative method to extrapolate KB integrals G^V to the thermodynamic limit is to use the scaling of LG^V with L, rather than the scaling of G^V with $1/L$. The scaling of G^V in Equation (2) can be rewritten as

$$LG^V(L) = G^\infty L + F^\infty \tag{12}$$

By fitting the linear part of the scaling of LG^V with L, it is possible to obtain G^∞ and F^∞. Finding the slope and intercepts of a straight line is easier than extrapolating the linear regime of the scaling of G^V with $1/L$. Another advantage of this approach is that an estimation of the surface effects is automatically computed. This estimation can be compared to other available methods for computing F^∞. So far, it is shown that three methods are available for estimating G^∞ from integrals of finite subvolumes:

1. Using the scaling of G^V (Equation (4)) with $1/L$. To estimate G^∞, the linear regime of the scaling is extrapolated to the limit $1/L \to 0$.
2. Using the direct extrapolation formula G_2 (Equation (6)) combined with the function $u_2(r)$ (Equation (7)). This will converge to G^∞ for large L.
3. Computing G^∞ from fitting the linear regime of the scaling of LG^V with L (Equation (12)). The values of the integrals G^V are computed using Equation (4).

To simplify the estimation of KB integrals, it would be useful to evaluate the performance of these methods in terms of accuracy and practicality. Similarly, different methods are available to compute the surface term in the thermodynamic limit F^∞:

1. Using the expression in Equation (11).
2. From the scaling of LF^V with L (Equation (9)). F^V is computed using Equation (10). The value of F^∞ is obtained from the slope of the scaling; as $LF^V(L) = F^\infty L + C$, in which C is a constant.
3. From the scaling of LG^V with L (Equation (9)). The value of F^∞ is obtained from the intercept of the scaling.

The objective of this work is to test the estimation of KB integrals G^∞ and the surface effects F^∞ using the approaches discussed earlier. For both G^∞ and F^∞, the effect of the size of the system is studied. These effects are investigated for both Lennard–Jones (LJ) and Weeks–Chandler–Andersen (WCA) fluids [21] at different densities. Finally, this work aims at quantifying the contributions of the surface term when computing KB integrals of LJ fluid at various densities.

This paper is organized as follows: In Section 2, the methods used to compute RDFs, KB integrals, and the surface term of KB integrals of LJ and WCA fluids are presented. Section 2 includes the details of the MD simulations. In Section 3, the results are presented, which include KB integrals and the surface term for WCA and LJ systems at different sizes and densities. Section 4 summarises the main findings of this work.

2. Methods

RDFs of systems of particles interacting via the LJ potential are computed using MD simulations in the NVT ensemble. Systems with different densities and number of particles are studied. Also, systems of particles interacting via the Weeks–Chandler–Andersen (WCA) potential [22], where only the repulsive part of the LJ potential is included, are considered. The common approach of particles counting was implemented to compute RDFs. While this is not carried out in this work, it is possible to investigate other methods. It would be interesting to see if force-based computations of RDFs improve the convergence of computed KB integrals [23–25]. For each system, the computed RDF is used to compute KB integrals G^∞ and the surface term F^∞ in the thermodynamic limit. For both quantities, the methods discussed in Section 1 are used. In this section, the numerical details of computing RDFs and the required integrals are briefly discussed.

According to Kirkwood–Buff theory, KB integrals are defined for open and infinite systems [6]. When computing KB integrals using molecular simulations of closed systems, it is essential to correct RDFs for finite-size effects [10,12,15]. Recently, a number of corrections for the RDFs have been proposed [12,26,27]. In Reference [15] it was demonstrated that the accuracy of computing KB integrals improves when the Ganguly and van der Vegt correction [26] is applied. Applying the Ganguly and van der Vegt correction results in RDFs that are consistent with the physical behavior of fluids. For example, Equation (13) converges to $g(r) = 1$ for a single-component ideal gas, which is the correct value in the thermodynamic limit. The Ganguly and van der Vegt [26] correction is based on the excess (or depletion) of the density of the system beyond a distance L from a central molecule α. The corrected RDF is

$$g_{\alpha\beta}^{\mathrm{vdV}}(r) = g_{\alpha\beta}(r) \frac{N_\beta \left(1 - \frac{V}{V_{\mathrm{box}}}\right)}{N_\beta \left(1 - \frac{V}{V_{\mathrm{box}}}\right) - \Delta N_{\alpha\beta}(r) - \delta_{\alpha\beta}} \tag{13}$$

$g_{\alpha\beta}(r)$ is obtained from a simulation in a finite system with total volume V_{box}. $\Delta N_{\alpha\beta}(r)$ is the excess number of particles of type β in a sphere of radius r around a particle of type α, which is computed by

$$\Delta N_{\alpha\beta}(r) = \int_0^r \mathrm{d}r' 4\pi r'^2 \rho_\beta \left[g_{\alpha\beta}(r') - 1\right] \tag{14}$$

For all systems studied in this work, RDFs are corrected using the Ganguly and van der Vegt corrections. The corrected RDFs are numerically integrated to obtain G^V, G_2, F^V, and F_2^∞. Once these quantities are obtained, various methods are implemented to estimate KB integrals G^∞ and the surface terms F^∞ in the thermodynamic limit. Table 1 provides the relations and description of the methods considered to estimate G^∞. Similarly, Table 2 presents information regarding the methods used to estimate F^∞.

Table 1. A brief description of the methods used in this work to estimate Kirkwood–Buff (KB) integrals in the thermodynamic limit G^∞ using radial distribution functions (RDFs) computed from finite systems.

Method	Equations	Description
1. Scaling of G^V with $1/L$	(4) and (5)	G^∞ is obtained from extrapolating the linear regime of the scaling to $1/L \to 0$.
2. Direct estimation G_2	(6) and (7)	A plateau in G_2 is identified when plotted as a function of L. To estimate G^∞, values of G_2 in this plateau are averaged.
3. Scaling of LG^V with L	(4) and (12)	To find G^∞, the slope of the linear part of the scaling is computed.

Table 2. A brief description of the methods used in this work to estimate the surface term in the thermodynamic limit F^∞ using RDFs computed from finite systems.

Method	Equations	Description
1. Direct estimation F_2^∞	(11)	A plateau in F_2^∞ is identified when plotted as a function of L. To estimate F^∞, values of F_2^∞ in this plateau are averaged.
2. Scaling of LF^V with $1/L$	(9) and (10)	To find F^∞, the slope of the linear part of the scaling is computed.
3. Scaling of LG^V with $1/L$	(4) and (12)	To find F^∞, the intercept of the linear part of the scaling is computed.

Simulation Details

RDFs of LJ and WCA fluids were computed using MD simulations and then used to estimate KB integrals and surface effects. The simulations were carried out using an in-house FORTRAN code. All RDFs were computed from simulations in the NVT ensemble. The systems were simulated at a dimensionless temperature $T = 2$, dimensionless densities ρ ranging from 0.2 to 0.8 and using number of particles N equals to 100, 500, 1000, 5000, 10,000, 30,000, and 50,000. For each size, the length of the simulation box L was set according to the required density.

All MD simulations started from a randomly generated configuration for which an energy minimization was used to eliminate particle overlaps. A sufficient number of time steps was used to initialize the system. After initialization, RDFs were sampled every 100 time steps. For both initialization and production, a dimensionless time step equal to 0.001 was used. The simulation length was chosen depending on the size of the system and the available computational resources. For instance, for systems with $N = 100$, 1×10^9 production time steps were carried out, while for the maximum size $N = 50,000$, 7×10^5 steps were used. Multiple independent simulations were performed for each point (ρ, N). The resulting RDFs were then averaged and used to compute G^∞ and F^∞. At high densities ($\rho > 0.4$), RDFs from at least 10 runs are used. At lower densities, at least 20 runs are performed to enhance statistics.

3. Results

3.1. Estimation of KB Integrals

KB integrals in the thermodynamic limit G^∞ are obtained using the three different approaches discussed earlier. To compare the estimation methods, WCA systems were studied while fixing temperature and density. These parameters define the thermodynamic state of the system. Values of KB integrals, computed using different methods, for other densities for LJ and WCA fluids are provided in the Supporting Information (SI). After comparing estimation methods of KB integrals, the relation between density of the system and KB integrals for LJ and WCA system is discussd.

Figure 1 shows RDFs for systems of different sizes of a WCA fluid at $T = 2$ and $\rho = 0.6$ (dimensionless units). Figure 1b shows that using small system sizes, specifically $N = 100$ and $N = 500$, results in RDFs with higher oscillations than large systems, where N equals to or larger than 1000. As will be shown later, this causes implications in the computation of G^∞. In Figure 2, the scaling of KB integrals of finite subvolumes G^V with $1/L$ is presented. For large systems, where $N > 500$, a linear range is identified which can be extrapolated to the limit $1/L \to 0$. Instead of computing G^V, KB integrals G^∞ can be directly estimated from RDFs using Equations (6) and (7). Figure 3 shows the estimation of G_2 for systems with varying sizes. When plotted as a function of L, the values of the integrals G_2 show a plateau at a constant value which corresponds to G^∞. However, Figure 3b shows that this is not true for all system sizes. In fact, the values of G^∞ can be accurately estimated for systems with a minimum number of particles of 5000, which is larger than the minimum size required in the previous extrapolation method (Figure 2). The third method to find G^∞ is to use the scaling of LG^V with L (Equation (2)). Figure 4 shows that plotting the integrals of finite subvolumes as LG^V vs. L results in a clear linear regime that is easily identified. The value of the slope of the fitted line corresponds to the value of G^∞. For this method, systems with number of particles equal to or larger than 500 can already be used to compute G^∞. In principle, all methods for estimating G^∞ should result in the same answer in the thermodynamic limit. In Table 3, values of KB integrals G^∞ obtained using the three methods studied in this work are listed. For KB integrals reported in Table 3, only uncertainties greater than 0.01% are shown. The values are obtained from systems with various sizes. For each size, a linear range was used to compute G^∞. In Reference [15], guidelines were provided for selecting a range for the extrapolation of G^V vs. $1/L$. Essentially, the first few molecular diameters after $r = 0$, and distances beyond $L/2$ should be avoided. Fitting lines of the scaling of LG^V vs. L is more convenient. In general, fitting regions are chosen such that the coefficient of determination R^2 is equal to or very close to 1. The values of G^∞ in Table 3 show that the three methods provide very similar estimations with statistical uncertainties below 0.1%. Moreover, results in Table 3 show that computing G^∞ using the direct estimation G_2 require larger systems compared to the other methods. This was found to be true for other densities as well as for systems with LJ particles (see the SI).

From studying other systems, it was found that the scaling of LG^V with L is found to be the easiest method to apply.

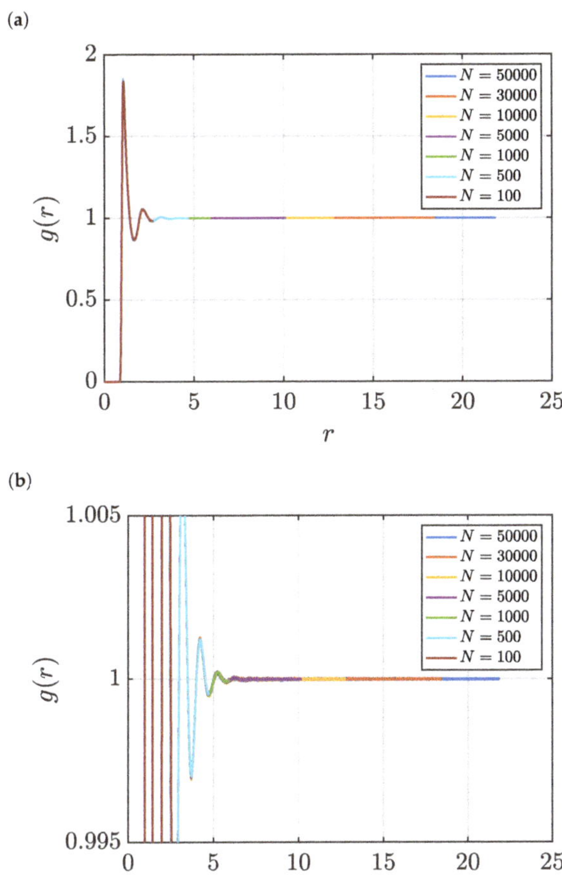

Figure 1. (a) RDFs for systems of different sizes of a Weeks–Chandler–Andersen (WCA) fluid at $T = 2$ and $\rho = 0.6$ (dimensionless units). Molecular dynamics (MD) simulations in the NVT ensemble were used to compute $g(r)$, and the Ganguly and van der Vegt correction [26] was applied (Equation (13)). (b) A zoom in of the plots in Figure (a) is shown.

Table 3. KB integrals in the thermodynamic limit G^∞ for a WCA system at $T = 2$ and $\rho = 0.6$ (dimensionless units). Values of G^∞ are computed from systems with various number of particles N and using the different methods listed in Table 1.

N	Scaling of G^V with $1/L$	Direct Estimation G_2	Scaling of LG^V with L
500	-1.5063 ± 0.0003	n/a	-1.5057 ± 0.0008
1000	-1.5027 ± 0.0000	n/a	-1.5028 ± 0.0002
5000	-1.5012 ± 0.0000	-1.5017 ± 0.0004	-1.5013 ± 0.0002
10,000	-1.5012 ± 0.0000	-1.5015 ± 0.0004	-1.5012 ± 0.0001
30,000	-1.5004 ± 0.0001	-1.5007 ± 0.0007	-1.5003 ± 0.0006
50,000	-1.4999 ± 0.0001	-1.5002 ± 0.0009	-1.500 ± 0.001

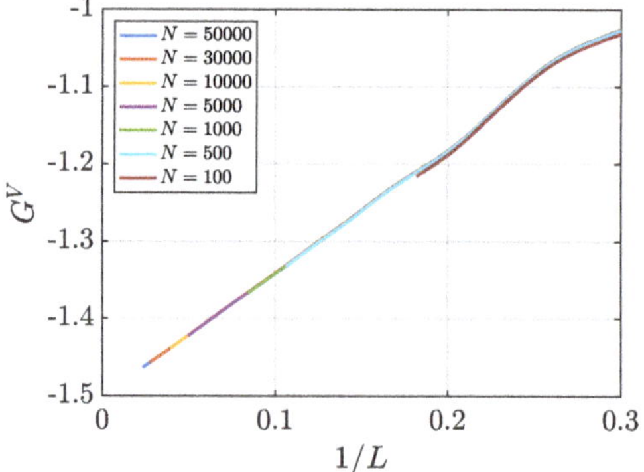

Figure 2. KB integrals of finite spherical subvolumes G^V (Equation (4)) vs. $1/L$ (L is the diameter) for the WCA fluid at $T = 2$ and $\rho = 0.6$ (dimensionless units). The values of G^V are computed for systems with varying number of molecules N. The used RDFs are provided in Figure 1.

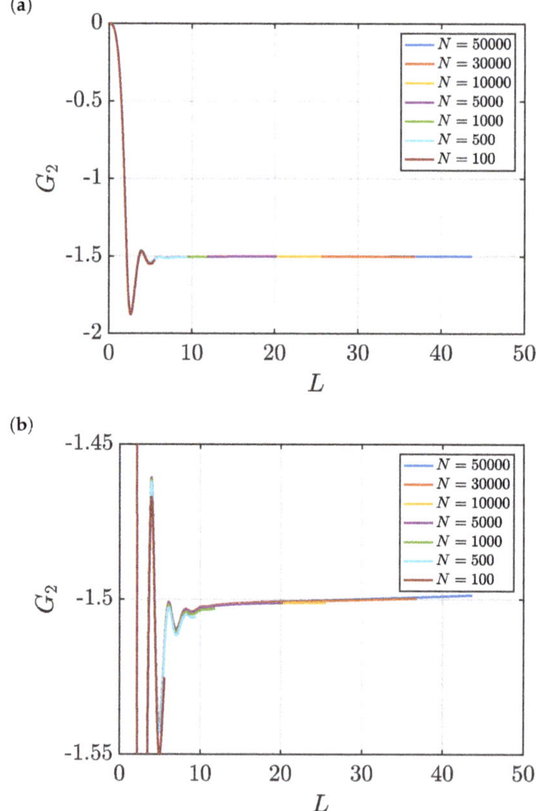

Figure 3. (a) Estimation of KB integrals in the thermodynamic limit, G_2 (Equations (6) and (7)) vs. L for the WCA fluid at $T = 2$ and $\rho = 0.6$ (dimensionless units). The values of G_2 are computed for systems with varying number of molecules N. The used RDFs are provided in Figure 1. (b) A zoom in of the plots in Figure (a) is shown.

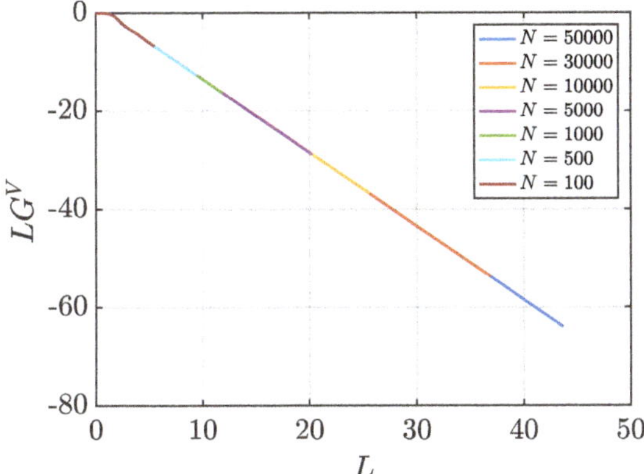

Figure 4. KB integrals of finite subvolumes multiplied by L, LG^V (Equation (4)) vs. L for a WCA fluid at $T = 2$ and $\rho = 0.6$ (dimensionless units). The values of G^V are computed for systems with varying number of molecules N. The used RDFs are provided in Figure 1.

The differences between the estimation methods can be further demonstrated by using a system of LJ particles at $\rho = 0.4$, which is more difficult to sample compared to the previously studied system. In Figure 5, RDFs of systems of varying sizes of a LJ fluid at $\rho = 0.4$ and $T = 2$ are shown. In Table 4, KB integrals G^∞ computed using the RDFs in Figure 5 are provided. The scaling of G^V with $1/L$ is shown in Figure 6. For this method, a linear range is not obtained for all system sizes. Systems with at least $N = 1000$ particles can be used to extrapolate to the thermodynamic limit. In Figure 7, KB integrals G_2 are plotted as a function of the size of the subvolume L. The figure shows that even larger systems are needed to find a reasonable estimate of G^∞ using G_2. Figure 7b shows that a plateau is only achieved for large systems where N equals to or larger than 5000. For this system (LJ fluid at $\rho = 0.4$), it is possible to use the scaling of LG^V with L to compute KB integrals from small sizes. Figure 8 demonstrates that straight lines that are easily fitted are achieved when using the scaling of LG^V with L, even for sizes where an estimation can not be made with the other two methods.

Table 4. KB integrals in the thermodynamic limit G^∞ for a LJ system at $T = 2$ and $\rho = 0.4$ (dimensionless units). Values of G^∞ are computed from systems with various number of particles N and using the different methods listed in Table 1.

N	Scaling of G^V with $1/L$	Direct Estimation G_2	Scaling of LG^V with L
500	n/a	n/a	-1.1593 ± 0.0001
1000	-1.1395 ± 0.0001	n/a	-1.1390 ± 0.0008
5000	-1.1161 ± 0.0006	-1.13 ± 0.02	-1.114 ± 0.004
10,000	-1.1156 ± 0.0005	-1.13 ± 0.02	-1.114 ± 0.004
30,000	-1.1064 ± 0.0009	-1.12 ± 0.02	-1.10 ± 0.01

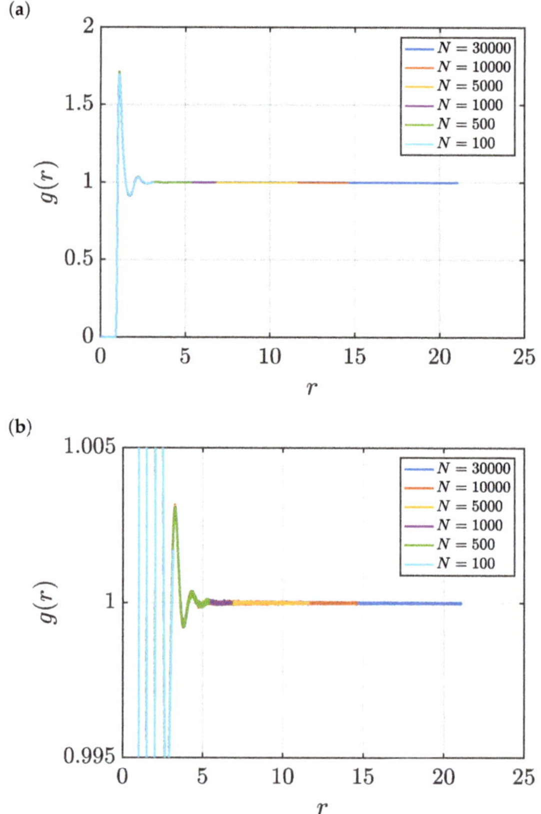

Figure 5. (a) RDFs for systems of different sizes of a LJ fluid at $T = 2$ and $\rho = 0.4$ (dimensionless units). MD simulations in the NVT ensemble were used to compute $g(r)$, and the Ganguly and van der Vegt correction [26] was applied (Equation (13)). (b) A zoom in of the plots in Figure (a) is shown.

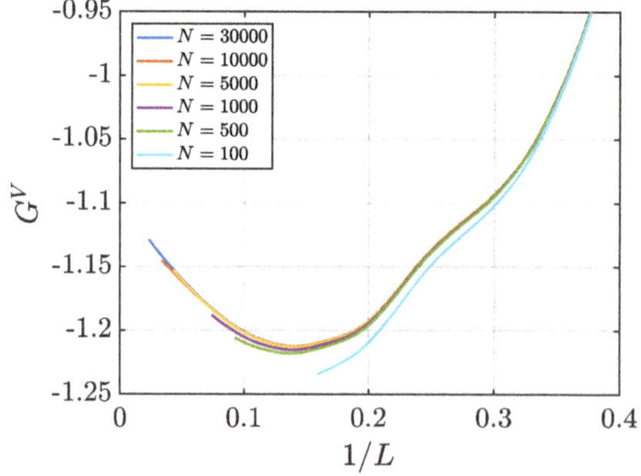

Figure 6. KB integrals of finite spherical subvolumes G^V (Equation (4)) vs. $1/L$ (L is the diameter) for the LJ fluid at $T = 2$ and $\rho = 0.4$ (dimensionless units). The values of G^V are computed for systems with varying number of molecules N. The used RDFs are provided in Figure 5.

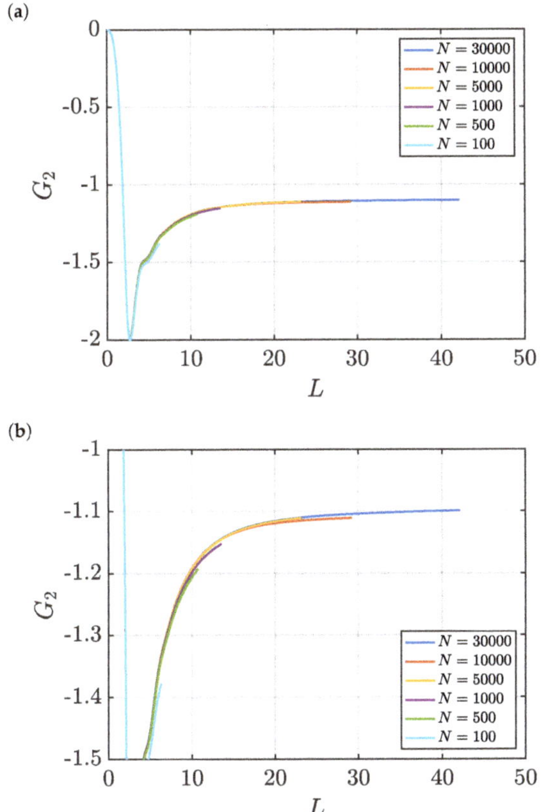

Figure 7. (**a**) Estimation of KB integrals in the thermodynamic limit, G_2 (Equations (6) and (7)) vs. L for the LJ fluid at $T = 2$ and $\rho = 0.4$ (dimensionless units). The values of G_2 are computed for systems with varying number of molecules N. The used RDFs are provided in Figure 5. (**b**) A zoom in of the plots in Figure (**a**) is shown.

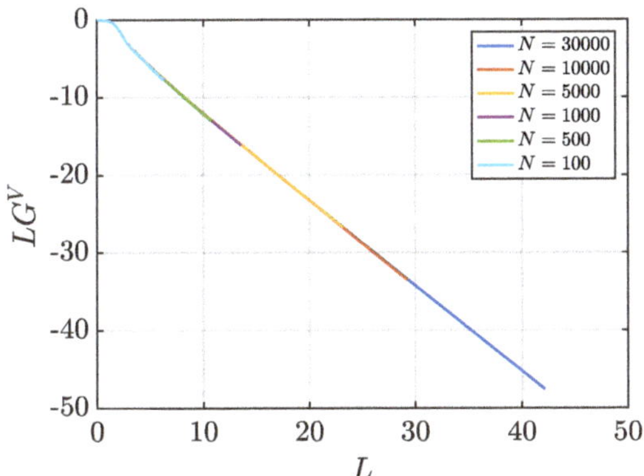

Figure 8. KB integrals of finite subvolumes multiplied by L, LG^V (Equation (4)) vs. L for a LJ fluid at $T = 2$ and $\rho = 0.4$ (dimensionless units). The values of G^V are computed for systems with varying number of molecules N. The used RDFs are provided in Figure 5.

Effect of System Size and Density

Figure 9a shows the effect of the size of the system on the values of G^∞ computed using the scaling of LG^V with L. The obtained values of G^∞ are practically constant. For the LJ fluid, a weak decrease, roughly linear in $N^{-1/3}$, is observed. Figure 9a, Tables 3 and 4 demonstrate that statistical uncertainties are small for systems with intermediate sizes ($N = 5000$ and $N = 10,000$). Smaller systems do not provide a sufficient linear regime and very large systems require longer sampling. In Figure 9b, KB integrals at different densities are shown for the LJ and WCA fluids. To estimate G^∞, the scaling of LG^V with L was used. MD simulations were performed to study systems with dimensionless densities ranging from 0.1 to 0.8.

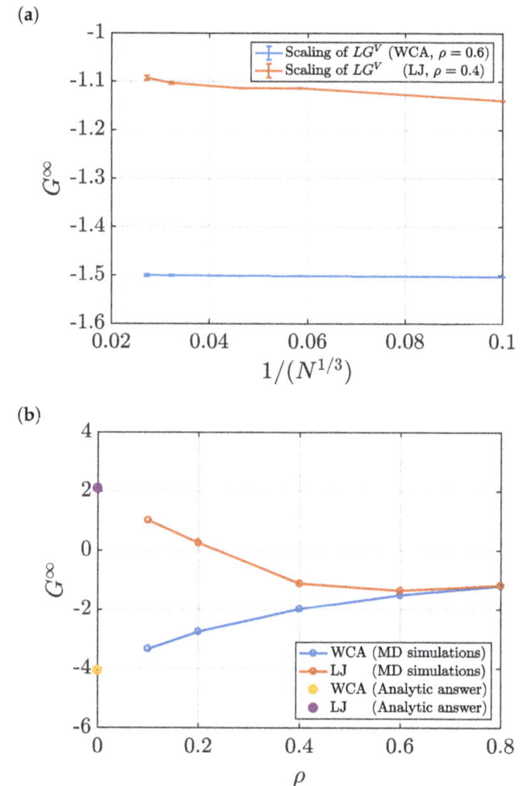

Figure 9. (a) KB integrals in the thermodynamic limit G^∞ as a function of the size of the system for the WCA fluid at $\rho = 0.6$, and the LJ fluid at $\rho = 0.4$. Both fluids are simulated at $T = 2$ (dimensionless units). (b) G^∞ as a function of dimensionless density ρ of LJ and WCA systems at $T = 2$. For all densities, the same number of particles is used, $N = 10,000$.

The behaviour of KB integrals in the limit $\rho \to 0$ can be checked by using the fact that in this limit, the RDF is known analytically, $g(r) = \exp[-\beta u(r)]$, where $u(r)$ is the pair potential [28]. Figure 9b shows that for both interaction potentials, the values of G^∞ computed using molecular simulation approach the correct value in the low density limit. In the high density limit, the differences between G^∞ of LJ and WCA fluids seem to disappear. At high densities, the repulsive part of the interaction potential, which is the same for WCA and LJ, becomes more important. Hence, the two fluids are expected to behave in the same way as the density increases. This is shown in Figure 9b.

3.2. Estimation of Surface Effects

An important objective of this work is to investigate surface effects of finite systems used to compute KB integrals. As mentioned earlier, there are three possible approaches to compute the surface term in the thermodynamic limit F^∞. Similar to the estimation of G^∞, the surface term of the WCA fluid is computed from systems with varying number of particles N at the same thermodynamic state.

In Figure 10, estimations of the surface term in the thermodynamic limit F_2^∞ (Equation (11)) are presented as a function of L. Unlike the values of G_2^∞, a plateau where the values can be averaged is not easily identified. Alternatively, it is possible to consider the scaling of the values of the surface term of finite subvolumes F^V (Equation (10)). Figure 11 shows the scaling of LF^V with L for the same WCA fluid. As in the case of the scaling of LG^V, a linear regime to be fitted is easily identified. The slope corresponds to the value of F^∞. Additionally, the value of F^∞ can be estimated from the intercept with the vertical axis of the line formed from the scaling of LG^V with L. The latter two approaches require smaller system sizes than the direct estimation F_2^∞. For instance, Figure 11 shows that systems with as few as $N = 1000$ provide a clear linear range that can be used to estimate F^∞. It is of interest to investigate whether the different available methods to find F^∞ result in matching estimations. The values of F^∞ computed using the three methods considered in this work are listed in Table 5. Results are shown for systems with varying number of particles. While acceptable statistics are achieved for most methods and system sizes, the values of F^∞ from the three different methods agree less well than the corresponding values of G^∞. This can be attributed to the larger statistical errors obtained when compared to estimating G^∞.

Table 5. Surface term in the thermodynamic limit F^∞ for a WCA system at $T = 2$ and $\rho = 0.6$ (dimensionless units). Values of F^∞ are computed from systems with various number of particles N and using the different methods listed in Table 2.

N	Direct Estimation F_2^∞	Scaling of LG^V with L	Scaling of LF^V with L
500	n/a	0.8168 ± 0.0008	n/a
1000	n/a	0.8082 ± 0.0002	0.804 ± 0.004
5000	0.801 ± 0.002	0.8036 ± 0.0002	0.8004 ± 0.0002
10,000	0.8013 ± 0.0004	0.8034 ± 0.0001	0.8013 ± 0.0003
30,000	0.795 ± 0.005	0.7979 ± 0.0006	0.79 ± 0.01
50,000	0.79 ± 0.01	0.793 ± 0.001	0.78 ± 0.02

As in the case of computing G^∞, using scaling of LG^V provides estimations of the surface term using systems smaller than those required by the other methods. This effect is more significant when looking at a system of LJ particles at a relatively low density. Table 6 provides the values of F^∞ for a LJ fluid at $\rho = 0.4$, computed using the methods studied in this work. In Figures 12 and 13 the scaling of F_2^∞ with L as well as the scaling of LF^V with L are shown. These plots illustrate that linear regions are not easily identified to compute F^∞, in contrast to the higher density case with $\rho = 0.6$ (Figures 10 and 11). As a result, when computing surface effects, the scaling of LG^V with L is recommended.

Table 6. Surface term in the thermodynamic limit F^∞ for a LJ system at $T = 2$ and $\rho = 0.4$ (dimensionless units). Values of F^∞ are computed from systems with various number of particles N and using the different methods listed in Table 2.

N	Direct Estimation F_2^∞	Scaling of LG^V with L	Scaling of LF^V with L
500	n/a	-0.2483 ± 0.0001	n/a
1000	n/a	-0.3320 ± 0.0008	n/a
5000	-0.53 ± 0.04	-0.460 ± 0.004	-0.5718 ± 0.0001
10,000	-0.52 ± 0.03	-0.464 ± 0.004	-0.5433 ± 0.0004
30,000	-0.60 ± 0.06	-0.543 ± 0.008	-0.6315 ± 0.0009

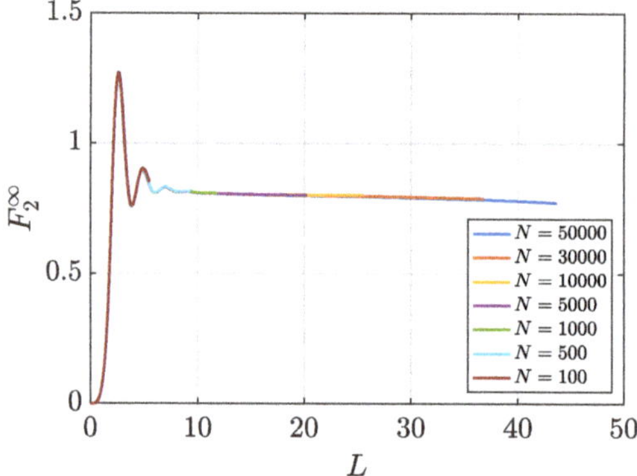

Figure 10. Estimation of the surface term in the thermodynamic limit F_2^∞ (Equation (11)) vs. L for the WCA fluid at $T = 2$ and $\rho = 0.6$ (dimensionless units). The values of F_2^∞ are computed for systems with varying number of molecules N. The used RDFs are provided in Figure 1.

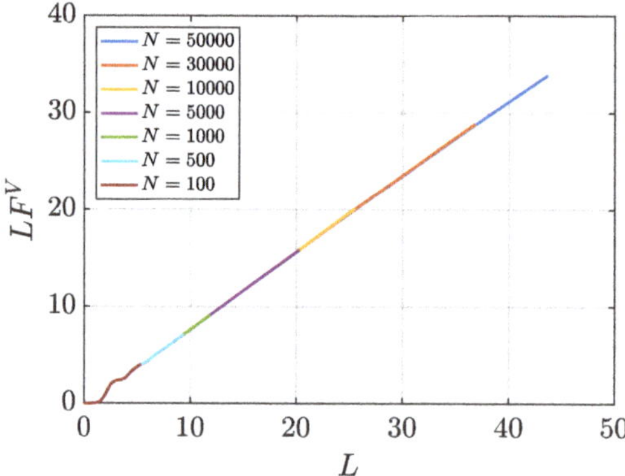

Figure 11. Surface effects of finite subvolumes multiplied by the diameter of the subvolume LF^v (Equation (10)) vs. L for the WCA fluid at $T = 2$ and $\rho = 0.6$ (dimensionless units). The values of G^v are computed for systems with varying number of molecules N. The used RDFs are provided in Figure 1.

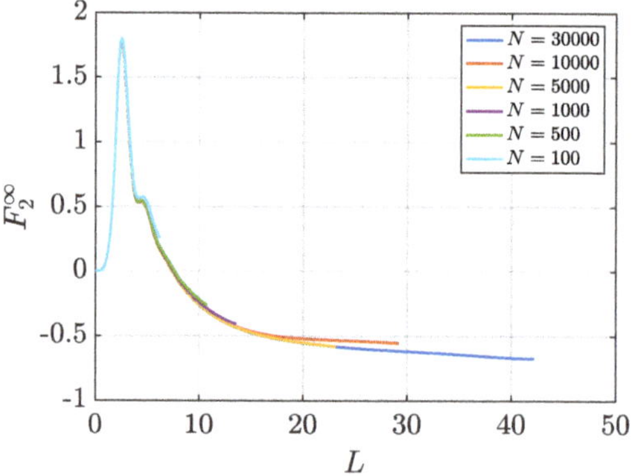

Figure 12. Estimation of the surface term in the thermodynamic limit F_2^∞ (Equation (11)) vs. L for the LJ fluid at $T = 2$ and $\rho = 0.4$ (dimensionless units). The values of F_2^∞ are computed for systems with varying number of molecules N. The used RDFs are provided in Figure 5.

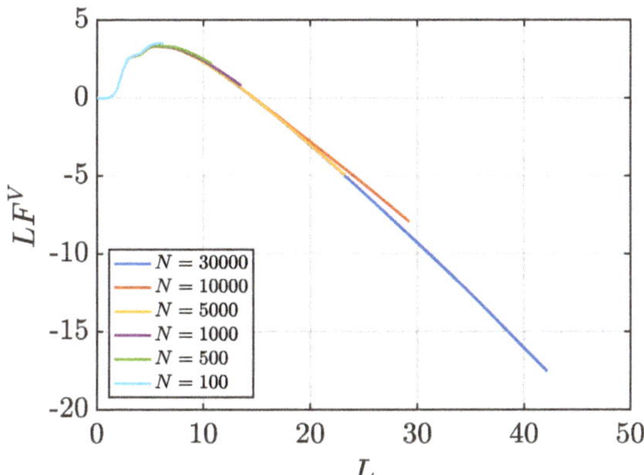

Figure 13. Surface effects of finite subvolumes multiplied by the diameter of the subvolume LF^V (Equation (10)) vs. L for the LJ fluid at $T = 2$ and $\rho = 0.4$ (dimensionless units). The values of G^V are computed for systems with varying number of molecules N. The used RDFs are provided in Figure 5.

Effect of System Size and Density

Figure 14a shows the dependence of F^∞ with inverse system size. Specifically, F^∞ decreases as $(1/N)^{1/3}$. This is observed for surface terms computed using the scaling of LF^V with L as well as values computed using the scaling of LG^V with L. In Figure 14a, the error bars of the values of F^∞ vary with N in a similar manner as the values of G^∞. Statistical uncertainties of the systems studied in this work are provided in Tables 5 and 6.

As for KB integrals, the surface terms F^∞ can be determined accurately as a function of the density. Figure 14b shows the values of F^∞ with density for LJ and WCA fluids. The surface term is estimated for the range $\rho = 0.1 - 0.8$. At $\rho \to 0$, F^∞ is computed analytically using Equation (10) and $g(r) = \exp[-\beta u(r)]$. In the low density limit, the surface terms computed in this work approach the theoretical value. In the high density limit, differences between surface term of LJ and WCA disappear due to dominating repulsive interactions, which are the same for the LJ and WCA potentials. From Figures 9b and 14b, a comparison between the values of G^∞ and F^∞ can be made. Both values change in the same manner with the density of the system. For all densities, surface terms F^∞ seem to have the same order of magnitude as KB integrals G^∞. This indicates the significant contribution of surface effects of finite systems used to compute KB integrals in the thermodynamic limit. The last observation applies to a LJ fluid as well as a WCA fluid.

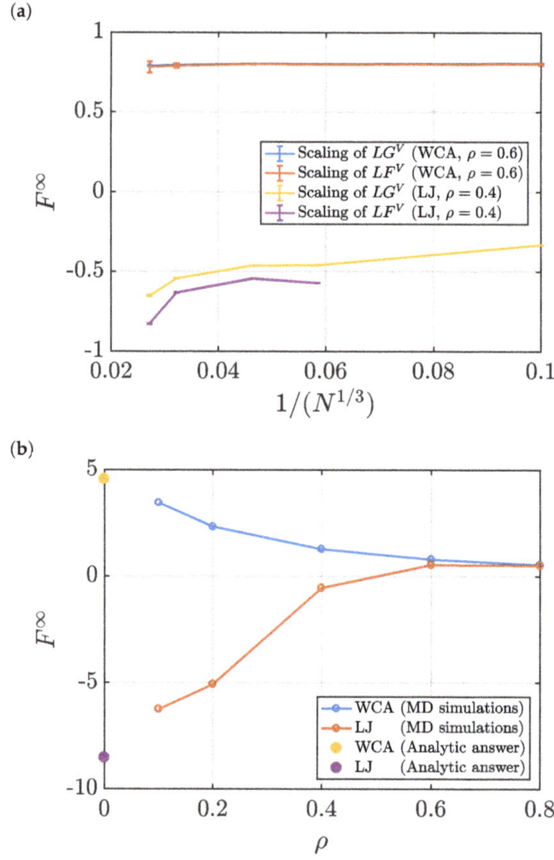

Figure 14. (**a**) Surface term in the thermodynamic limit F^∞ as a function of the size of the system for the WCA fluid at $\rho = 0.6$, and the LJ fluid at $\rho = 0.4$. Both fluids are simulated at $T = 2$ (dimensionless units). (**b**) F^∞ as a function of dimensionless density ρ of LJ and WCA systems at $T = 2$. For all densities, the same number of particles is used, $N = 10{,}000$.

4. Conclusions

In this work, KB integrals and surface effects in the thermodynamic limit were computed for systems of LJ and WCA fluids. Using MD simulations, RDFs of the LJ and WCA systems of different sizes were computed. Different methods were used to estimate KB integrals G^∞ from RDFs of finite systems: scaling of G^V with $1/L$, direct estimation integrals G_2, and the scaling of LG^V with L. The three methods were found to provide reliable estimates of G^∞. Differences between the three methods mainly arise from the size of the system required to obtain an accurate estimation. The scaling of LG^V with L was found to require smaller systems when compared to other methods. Moreover, the scaling of LG^V was found the easiest to implement for estimating KB integrals and it provides a suitable estimate of surface effects. Estimating the surface term in the thermodynamic limit F^∞ is possible from: the finite integral F_2^∞, the scaling of LF^V with L, as well as the scaling of LG^V with L. For all methods, the surface term F^∞ was found to decrease with increasing system size. The magnitude of the values of F^∞ were found to be the same as the magnitude of the KB integrals G^∞. Both quantities were found to change in the same manner with the density of the system. KB integrals and surface terms were computed for LJ and WCA fluids at different densities. The differences between KB integrals of the two systems, LJ and WCA, vanish for high densities as the structure is dominated by repulsive interactions.

Supplementary Materials: The following are available at http://www.mdpi.com/2079-4991/10/4/771/s1, Tables S1–S16: KB integrals G^{∞} and the surface term F^{∞} in the thermodynamic limit are provided at different densities for the LJ and WCA fluids. For each interaction potential, G^{∞} and F^{∞} are computed at a dimensionless temperature $T = 2$ and the following dimensionless densities: $\rho = 0.2$, $\rho = 0.4$, $\rho = 0.6$, and $\rho = 0.8$.

Author Contributions: Conceptualization, P.K., J.-M.S. and T.J.H.V.; investigation, all authors; methodology, P.K., J.-M.S., T.J.H.V. and S.K.S.; supervision: P.K., J.-M.S., O.A.M., T.J.H.V., S.K.S. and I.G.E.; writing–review and editing: all authors; writing–original draft preparation: N.D. All authors have read and agreed to the published version of the manuscript.

Funding: This work was sponsored by NWO Exacte Wetenschappen (Physical Sciences) for the use of supercomputer facilities with financial support from the Nederlandse Organisatie voor weten schappelijk Onderzoek (Netherlands Organization for Scientific research, NWO). TJHV acknowledges NWO-CW for a VICI grant. PK acknowledges funding by JSPS KAKENHI Grant Number 19K05383.

Conflicts of Interest: The authors declare no conflict of interest.

References

1. Patterson, D. Effects of molecular size and shape in solution thermodynamics. *Pure Appl. Chem.* **1976**, *47*, 305–314. [CrossRef]

2. Koga, Y. *Solution Thermodynamics and Its Application to Aqueous Solutions: A Differential Approach*, 2nd ed.; Elsevier: Amsterdam, The Netherlands, 2017.

3. Lee, L.L. *Molecular Thermodynamics of Non-Ideal Fluids*; Butterworth-Heinemann: Oxford, UK, 2016.

4. Panagiotopoulos, A.Z. Molecular simulation of phase equilibria: Simple, ionic and polymeric fluids. *Fluid Phase Equilib.* **1992**, *76*, 97–112. [CrossRef]

5. Schnell, S.K.; Skorpa, R.; Bedeaux, D.; Kjelstrup, S.; Vlugt, T.J.H.; Simon, J.M. Partial molar enthalpies and reaction enthalpies from equilibrium molecular dynamics simulation. *J. Chem. Phys.* **2014**, *141*, 144501. [CrossRef]

6. Kirkwood, J.G.; Buff, F.P. The Statistical Mechanical Theory of Solutions. I. *J. Chem. Phys.* **1951**, *19*, 774–777. [CrossRef]

7. Ben-Naim, A. *Molecular Theory of Solutions*; Oxford University Press: Oxford, UK, 2006.

8. Hall, D.G. Kirkwood–Buff theory of solutions. An alternative derivation of part of it and some applications. *Trans. Faraday Soc.* **1971**, *67*, 2516–2524. [CrossRef]

9. Newman, K.E. Kirkwood–Buff solution theory: Derivation and applications. *Chem. Soc. Rev.* **1994**, *23*, 31–40. [CrossRef]

10. Dawass, N.; Krüger, P.; Schnell, S.K.; Simon, J.M.; Vlugt, T.J.H. Kirkwood-Buff integrals from molecular simulation. *Fluid Phase Equilib.* **2019**, *486*, 21–36. [CrossRef]

11. Schnell, S.K.; Liu, X.; Simon, J.M.; Bardow, A.; Bedeaux, D.; Vlugt, T.J.H.; Kjelstrup, S. Calculating thermodynamic properties from fluctuations at small scales. *J. Phys. Chem. B* **2011**, *115*, 10911–10918. [CrossRef]

12. Krüger, P.; Schnell, S.K.; Bedeaux, D.; Kjelstrup, S.; Vlugt, T.J.H.; Simon, J.M. Kirkwood–Buff Integrals for Finite Volumes. *J. Phys. Chem. Lett.* **2013**, *4*, 235–238. [CrossRef]

13. Hill, T.L. Thermodynamics of Small Systems. *J. Chem. Phys.* **1962**, *36*, 3182–3197. [CrossRef]

14. Hill, T.L. *Thermodynamics of Small Systems*; Dover: New York, NY, USA, 1994.

15. Dawass, N.; Krüger, P.; Schnell, S.K.; Bedeaux, D.; Kjelstrup, S.; Simon, J.M.; Vlugt, T.J.H. Finite-size effects of Kirkwood–Buff integrals from molecular simulations. *Mol. Simu.* **2018**, *44*, 599–612. [CrossRef]

16. Krüger, P.; Vlugt, T.J.H. Size and shape dependence of finite–volume Kirkwood-Buff integrals. *Phys. Rev. E* **2018**, *97*, 051301. [CrossRef] [PubMed]

17. Strøm, B.A.; Simon, J.M.; Schnell, S.K.; Kjelstrup, S.; He, J.; Bedeaux, D. Size and shape effects on the thermodynamic properties of nanoscale volumes of water. *Phys. Chem. Chem. Phys.* **2017**, *19*, 9016–9027. [CrossRef] [PubMed]

18. Frenkel, D.; Smit, B. *Understanding Molecular Simulation: From Algorithms to Applications*, 2nd ed.; Academic Press: London, UK, 2002; Volume 1.
19. Dawass, N.; Krüger, P.; Schnell, S.K.; Simon, J.M.; Vlugt, T.J.H. Kirkwood–Buff integrals of finite systems: Shape effects. *Mol. Phys.* **2018**, *116*, 1573–1580. [CrossRef]
20. Santos, A. Finite-size estimates of Kirkwood-Buff and similar integrals. *Phys. Rev. E* **2018**, *98*, 063302. [CrossRef]
21. Lennard-Jones, J.E. On the determination of molecular fields. II. From the equation of state of gas. *Proc. R. Soc. A* **1924**, *106*, 463–477.
22. Weeks, J.D.; Chandler, D.; Andersen, H.C. Role of Repulsive Forces in Determining the Equilibrium Structure of Simple Liquids. *J. Chem. Phys.* **1971**, *54*, 5237–5247. [CrossRef]
23. Borgis, D.; Assaraf, R.; Rotenberg, B.; Vuilleumier, R. Computation of pair distribution functions and three-dimensional densities with a reduced variance principle. *Mol. Phys.* **2013**, *111*, 3486–3492. [CrossRef]
24. De las Heras, D.; Schmidt, M. Better than counting: Density profiles from force sampling. *Phys. Rev. Lett.* **2018**, *120*, 218001. [CrossRef]
25. Jamali, S. Transport Properties of Fluids: Methodology and Force Field Improvement Using Molecular Dynamics Simulations. Ph.D. Thesis, Delft University of Technology, Delft, The Netherlands, 2020.
26. Ganguly, P.; van der Vegt, N.F.A. Convergence of Sampling Kirkwood-Buff Integrals of Aqueous Solutions with Molecular Dynamics Simulations. *J. Chem. Theory Comput.* **2013**, *9*, 1347–1355. [CrossRef]
27. Cortes-Huerto, R.; Kremer, K.; Potestio, R. Communication: Kirkwood-Buff integrals in the thermodynamic limit from small-sized molecular dynamics simulations. *J. Chem. Phys.* **2016**, *145*, 141103. [CrossRef]
28. Hansen, J.P.; McDonald, I.R. *Theory of Simple Liquids*; Elsevier: Amsterdam, The Netherlands, 1990.

 *nanomaterials*

Article

Characterizing Polymer Hydration Shell Compressibilities with the Small-System Method

Madhusmita Tripathy †, Swaminath Bharadwaj †, Shadrack Jabes B. † and Nico F. A. van der Vegt *

Eduard-Zintl-Institut für Anorganische und Physikalische Chemie, Technische Universität Darmstadt, 64287 Darmstadt, Germany; tripathy@cpc.tu-darmstadt.de (M.T.); bharadwaj@cpc.tu-darmstadt.de (S.B.); barnabas@cpc.tu-darmstadt.de (S.J.B.)
* Correspondence: vandervegt@cpc.tu-darmstadt.de
† These authors contributed equally to this work.

Received: 24 June 2020; Accepted: 22 July 2020; Published: 25 July 2020

Abstract: The small-system method (SSM) exploits the unique feature of finite-sized open systems, whose thermodynamic quantities scale with the inverse system size. This scaling enables the calculation of properties in the thermodynamic limit of macroscopic systems based on computer simulations of finite-sized systems. We herein extend the SSM to characterize the hydration shell compressibility of a generic hydrophobic polymer in water. By systematically increasing the strength of polymer-water repulsion, we find that the excess inverse thermodynamic correction factor $(\Delta 1/\Gamma_s^{\infty})$ and compressibility $(\Delta\chi_s)$ of the first hydration shell change sign from negative to positive. This occurs with a concurrent decrease in water hydrogen bonding and local tetrahedral order of the hydration shell water. The crossover lengthscale corresponds to an effective polymer bead diameter of 0.7 nm and is consistent with previous works on hydration of small and large hydrophobic solutes. The crossover lengthscale in polymer hydration shell compressibility, herein identified with the SSM approach, relates to hydrophobic interactions and macromolecular conformational equilibria in aqueous solution. The SSM approach may further be applied to study thermodynamic properties of polymer solvation shells in mixed solvents.

Keywords: small system method; thermodynamics of small systems; hydration shell thermodynamics; finite size correction

1. Introduction

Although the laws of statistical mechanics are customarily used in computer simulations to relate thermodynamical properties of a broad range of systems to interactions between their constituent components, there remains a caveat: computer simulations are performed on finite-sized systems in which the thermodynamic properties may deviate from the corresponding properties in the thermodynamic limit (TL) of large systems [1,2]. An example is the incorrect asymptotic behavior of the radial distribution function (RDF), $g(r)$, obtained by computer simulation of a closed system with a finite number of particles. Routes for calculating thermodynamic properties from integrals that involve the RDF are sensitive to subtle changes in the asymptotic tail behavior of the RDF and, therefore, show size dependencies in practical calculations. This issue is well known [3–5] and has received renewed attention in recent years, in particular, in applications of the Kirkwood-Buff theory of solutions and the calculation of Kirkwood-Buff integrals (KBI) [4–9].

The small-system method (SSM) by Schnell et al. [10,11] provides another route for computing thermodynamic properties in the TL from small scale fluctuations. This method uses a small-system scaling relation, derived based on the thermodynamics of small systems by Hill [12,13], to obtain thermodynamic properties corresponding to the TL. For example, the scaling relation for the thermodynamic correction factor, Γ, of a single-component small system is given by [10]

$$1/\Gamma(L) = 1/\Gamma^\infty + \frac{c}{L} \tag{1}$$

where L is the linear dimension of a small system with volume $V = L^3$, c is a constant that does not depend on L, and Γ^∞ is the thermodynamic correction factor in the TL where $V \to \infty$, $\langle N \rangle \to \infty$, and the particle number density $\rho = \langle N \rangle / V$ is constant. Γ^∞ is related to the isothermal compressibility (χ_T) according to

$$1/\Gamma^\infty = \rho k_{\mathrm{B}} T \chi_T. \tag{2}$$

The inverse thermodynamic correction factor in Equation (1) follows from density fluctuations in the grand-canonical ensemble

$$1/\Gamma(L) = \frac{\langle N^2 \rangle - \langle N \rangle^2}{\langle N \rangle}, \tag{3}$$

where N is the number of particles and $\langle \cdot \rangle$ indicates ensemble averages. In practice, $1/\Gamma(L)$ is calculated for small open systems of varying linear dimension L embedded in a large particle reservoir (closed simulation box of volume $V_b >> L^3$). The scaling relation (Equation (1)) is then observed for small-system sizes that obey $V_b^{1/3} > L > \xi$ (where ξ is the correlation length). The SSM has been successfully applied to pure component systems as well as binary and multicomponent systems [5,11,14–16]. Very recently, the method has been further employed to study the thermodynamic properties of confined fluids in nanopores [17,18]. In a spirit similar to the SSM, Jamali et al. [19] exploited the scaling of the elements of the diffusivity matrix, which depends on Γ, with system size to estimate their TL value for ternary molecular and model mixtures.

Recently, Trinh et al. [20,21] have extended the SSM to study the thermodynamic properties of adsorbed CO_2 and CH_4 layers on graphite and activated carbon surfaces. Motivated by these studies, we herein extend the SSM to investigate the compressibility of polymer hydration shells. The TL $(1/L \to 0)$ of an extended hydration shell will be considered for a fully stretched linear chain with L defined along the direction of the chain. The hydration shell compressibility of hydrophobic polymers relates to hydrophobic interactions and pressure dependencies of polymer (coil-to-globule) collapse equibria in water. We are, in particular, interested in the role of a crossover length scale in hydrophobic hydration of macromolecules. While hydration thermodynamics of small hydrophobic solutes is governed by density fluctuations in pure water, hydration of large hydrophobic solutes is governed by the (de)wetting of their surfaces [22,23]. The crossover length scale, above which a macroscopic thermodynamic description applies, has been investigated based on computer simulations that studied the hydration thermodynamics of small and large hydrophobic solutes [23–26]. Susceptibilities of extended hydration shells to external perturbations have been characterized more recently [27,28] and further demonstrates the role of dewetting of hydrophobic surfaces in the thermodynamics of biomolecular interactions such as protein-protein and ligand-protein interactions.

We will herein ask if a small-to-large crossover can be observed in the polymer hydration shell compressibility by systematically increasing the range of the repulsive interaction between beads of a generic polymer and water. Our results demonstrate that $\Delta 1/\Gamma_s^\infty$ and $\Delta \chi_s$, i.e., the inverse thermodynamic correction factor and the compressibility of the first hydration shell in excess to the corresponding properties of a shell of exactly the same radial and lateral dimensions in pure water, exhibit a crossover from negative to positive values at intermediate strengths of the polymer bead-water repulsion. The observed crossover corresponds to an effective polymer bead Lennard-Jones diameter of approximately 0.7 nm. This crossover length scale in hydrophobic hydration agrees with earlier work on hydrophobic hydration of spherical solutes in which a relation to the Egelstaff–Widom lengthscale, i.e., the product of the surface tension and compressibility, of water was demonstrated [29]. In the present work, we observe this small-to-large crossover for all lateral dimensions L of the small system (first hydration shell). However, $\Delta 1/\Gamma_s(L)$ is independent of L for effective polymer bead sizes below 0.7 nm, while it increases with L for hydrophobic polymers with larger effective bead sizes. This indicates that only for effective polymer bead sizes larger than 0.7 nm, the overall exposed hydrophobic polymer surface area impacts the excess thermodynamic properties of hydrophobic polymer hydration shells and causes a dependence of these properties on the linear polymer dimension whose limiting values can be determined with the small system method.

The remainder of this paper is structured as follows. In Section 2, we discuss the details of the simulations. In Section 3, we discuss the methodology used to estimate the compressibility of individual hydration shells and the results we obtain. We comment on the scope and limitations of the SSM approach in Section 4 and conclude in Section 5 with our main results and future directions.

2. Methods

In this work, a generic hydrophobic polymer model, developed by Zangi et al.[30], was used to simulate aqueous polymer systems at various strengths of the polymer-water repulsive interaction. The polymer chain consisted of 40 uncharged Lennard-Jones (LJ) beads (σ_p = 0.4 nm and ϵ_p = 1.0 kJ mol^{-1}) connected via rigid bonds of length of 0.153 nm. The simulation system was composed of the 40-mer chain, which was periodically replicated in the Z-direction to obtain an infinitely long polymer chain [31], along with 5000 TIP4P/2005 water [32] molecules. The box size along the Z-direction, l_z, was equal to the end-to-end distance R_{ee} of the 40-mer chain, which was fixed at 6.11 nm in all simulations to ensure a fully stretched polymer conformation. Thus, unlike in the original model [30], the equilibrium bond angles were kept at 180°. Additionally, position restraints, involving a harmonic potential with a force constant of 10^5 kJ mol^{-1}nm^{-2}, were applied on each polymer bead to keep the chain conformation fixed.

The MD simulations were performed using the Gromacs 2019.3 [33] package. All bonds were constrained using the LINCS algorithm [34]. For the van der Waals interaction, a cutoff of 1.4 nm was used without long range dispersion corrections. Long range electrostatic interactions were calculated using the particle mesh Ewald method [35] with a real space cutoff of 1.4 nm, a grid space of 0.12 nm, and an interpolation order of 10^{-4}. The pressure and temperature were fixed at 1 bar and 300 K, respectively. The simulation systems were energy minimized using the steepest descent method until convergence and subsequently equilibrated in the NVT ensemble for 1 ns using the velocity-rescale thermostat [36]. This was followed by a 1 ns simulation run under NPT conditions using the Nośe-Hoover thermostat [37], with τ_T =1.0 ps, and the Berendsen barostat [38], with τ_P = 2.0 ps. In the NPT simulations, the end-to-end distance of the chain ($R_{ee} = l_z$) was kept constant by applying a semi-isotropic pressure coupling, which allowed box fluctuations only in the X- and Y-directions. Subsequently, a 2 ns NPT run was performed using the Nośe-Hoover thermostat [37] (τ_T = 1.0 ps) and the Parinello–Rahman barostat [39] (τ_P = 2.0 ps). This NPT run was followed by 5 cycles of NPT (using Nośe-Hoover thermostat and Parinello–Rahman barostat) and NVT (using Nośe-Hoover thermostat) runs of 1 ns duration each in order to ensure that the pressure in the NVT simulations fluctuates around 1 bar. The last NVT simulation was extended by 10 ns to generate the production trajectory, which was used for subsequent analysis. For all simulations, an integration time-step of 2 fs was used, and the system snapshots were collected every 1 ps.

The polymer-water interaction was modeled using an LJ 12-6 potential with the form,

$$U_{pw} = \alpha \frac{C_{12,pw}}{r_{pw}^{12}} - \frac{C_{6,pw}}{r_{pw}^{6}}, \tag{4}$$

where $C_{12,pw} (= 4\epsilon_{pw}\sigma_{pw}^{12})$ and $C_{6,pw} (= 4\epsilon_{pw}\sigma_{pw}^{6})$ are, respectively, the repulsive and attractive contributions to the potential and α is a scaling factor for the repulsive interaction. The polymer-water interaction parameters, ϵ_{pw} and σ_{pw}, were calculated using Lorenz–Bertholet mixing rules. The repulsive part of the potential was systematically tuned such that at low repulsion (low α) the polymer resembles an infinite string of overlapping small hydrophobic beads while at high repulsion (high α) it resembles an extended hydrophobic surface. The simulations were performed with $\alpha = 1$, 2, 4, 6, 8, 10, 12, 15, and 20. As shown in Figure 1, an increase in α led to an increase in the minimal distance of approach between the polymer and water, and a corresponding decrease in their attractive interaction strength. It is, therefore, possible to calculate the effective size of the polymer beads ($\sigma_{\mathrm{p}}^{\mathrm{eff}}$) in terms of the interaction parameters and α:

$$
U_{\mathrm{pw}} = \alpha \frac{C_{12,\mathrm{pw}}}{r_{\mathrm{pw}}^{12}} - \frac{C_{6,\mathrm{pw}}}{r_{\mathrm{pw}}^{6}} = \frac{C_{12,\mathrm{pw}}^{\mathrm{eff}}}{r_{\mathrm{pw}}^{12}} - \frac{C_{6,\mathrm{pw}}^{\mathrm{eff}}}{r_{\mathrm{pw}}^{6}},
$$
$$
\sigma_{\mathrm{pw}}^{\mathrm{eff}} = \alpha^{1/6}\sigma_{\mathrm{pw}},
$$
$$
\sigma_{\mathrm{p}}^{\mathrm{eff}} = 2\sigma_{\mathrm{pw}}^{\mathrm{eff}} - \sigma_{\mathrm{ww}}.
$$

(5)

The inset in Figure 1 shows the change in $\sigma_{\mathrm{p}}^{\mathrm{eff}}$ as a function of α. $\sigma_{\mathrm{p}}^{\mathrm{eff}}$ increased with increasing α, approaching a value of 1 nm at $\alpha = 40$.

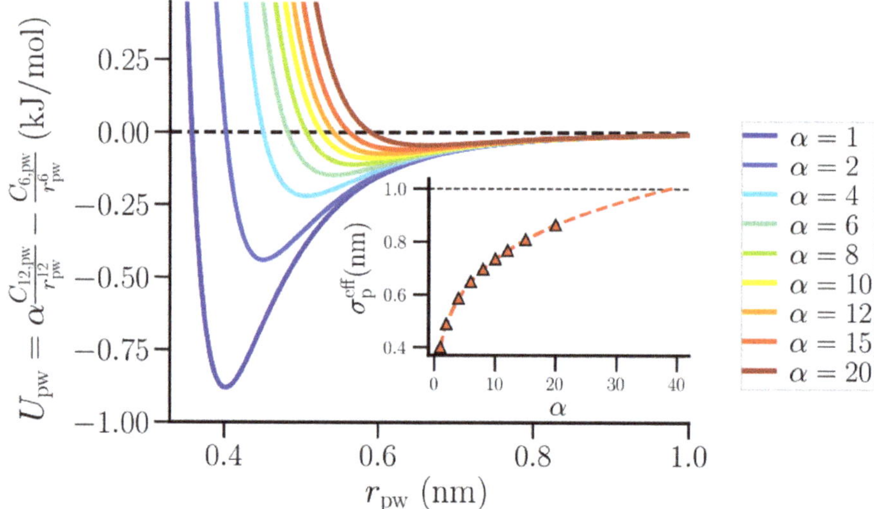

Figure 1. The polymer-water interaction potential, U_{pw}, at various repulsive interaction strength, α. The inset shows the relation between the effective size of the polymer beads, $\sigma_{\mathrm{p}}^{\mathrm{eff}}$, and α.

A pure water system, which was used as the reference for the analysis of hydration shell properties, consisted of 8000 TIP4P/2005 water molecules in a cubic simulation box. This system was equilibrated following the same protocol that we used for the aqueous polymer system, using repeated NPT-NVT equilibration cycles. This was done to ensure that the density of the system in the NPT ensemble equilibrates and the pressure in the NVT production trajectory fluctuates around 1 bar and well represents the ambient conditions. Unless and otherwise specified, all reference analyses for hydration shell properties were performed using this simulation.

As we are interested in the properties of the first few polymer hydration shells, the small systems were chosen to be cylindrical shells around the linearly extended polymer chain. This way, only the direct effect of polymer-water interaction is included in the analysis while avoiding any effect of polymer conformational fluctuations. There have been previous studies where small systems of various shapes were used to sample thermodynamic properties, which were found to scale identically with the surface to volume ratio of the small systems [15]. For all these shapes (sphere, cube, and polygons), the size of the small system was measured in terms of the diameter or edge length. On the other hand, a cylindrical shape of the small system involves two length parameters, the radius and the height. Therefore, to test our current approach involving cylindrical observation volumes, we first used it to estimate the bulk isothermal compressibility of SPC/E water [40]. The details of the analyses on bulk SPC/E water and aqueous polymer systems are discussed in the next section along with the corresponding results.

In addition to $\Delta 1/\Gamma_s^{\infty}$ and $\Delta \chi_s$ of the polymer hydration shells, changes in the structural order of the water molecules in the first hydration shell of the hydrophobic polymer were quantified in terms of the tetrahedral order parameter q_{tet}. For a central oxygen atom i surrounded by the nearest neighbor oxygen atoms $\{j,k\}$, q_{tet} of the central oxygen was calculated as [41,42]:

$$q_{\text{tet},i} = 1 - \frac{3}{8} \sum_{j=1}^{3} \sum_{k=j+1}^{4} (\cos \psi_{ijk} + 1/3)^2, \tag{6}$$

where ψ_{ijk} is the angle between the bond vectors $\mathbf{r}_{ij}$ and $\mathbf{r}_{ik}$. As the water molecules on the outer surface of the first hydration shell can have hydrogen bond neighbors in the second shell, those instances were included in the calculations to compute q_{tet} for the first shell as a function of α. For comparison, q_{tet} for shells in pure water were also computed using the polymer hydration shell widths corresponding to various α values.

In all cases, the 10 ns long production trajectory was divided into four windows of 2.5 ns each and the thermodynamic properties were calculated over the windows. The resulting data was averaged to get the mean values of the quantities along with the corresponding standard error.

3. Results and Discussion

3.1. Isothermal Compressibility of SPC/E Water: Sampling Fluctuations in Cylindrical Observation Volumes

A system of 8000 SPC/E water molecules was simulated (following the protocol discussed above) at 300 K and 360 K to benchmark the small box method using cylindrical shapes of the subvolumes. A point close to the center of the simulation box (with box size ~ 6.23 nm) was identified through which the reference axis was fixed along the Z-direction. Around this reference axis, particle number fluctuations were sampled in cylindrical observation volumes using the oxygen centers of the water molecules (Figure 2a). The radius of the cylinder (r_c) was varied between 0.25 nm to 2.5 nm. For a fixed value of r_c, the height of the cylinder (L) was varied in steps of 0.1 nm, starting from $L = 0.1$ nm (the smallest subvolume). Here, r_c and L together represented a small subvolume. For a given r_c, the observation cylinder was translated along the fixed axis, also in steps of 0.1 nm, for a given value of L and fluctuations were sampled. This led to improved statistics, especially for the values of L, which are less than half the box size. For a given r_c, $1/\Gamma(L, r_c)$ was obtained from the observed particle number fluctuations (Equation (3)). In the resulting data for $1/\Gamma(L, r_c)$ vs. $1/L$ (Figure 3a), a linear regime was identified between $L=1$ nm and 2 nm across all values of r_c. This linear regime was extrapolated to $1/L \to 0$, following Equation (1), to obtain $1/\Gamma(r_c)$, which characterizes particle number fluctuations in an infinitely long cylinder of radius r_c in bulk water. It should be noted that the simulation box behaves like a particle bath for the open small subvolumes. As the size of the subvolume becomes comparable to that of the simulation box, this assumption breaks down. Thus, data for large L values suffer from finite size effects and insufficient statistics. On the other hand, for very small values of L corresponding to a few particle diameters, $1/\Gamma(L, r_c)$ deviates from the scaling relation (Equation (1)) as well. Therefore, the data for intermediate values of L were considered for estimating the limiting bulk properties.

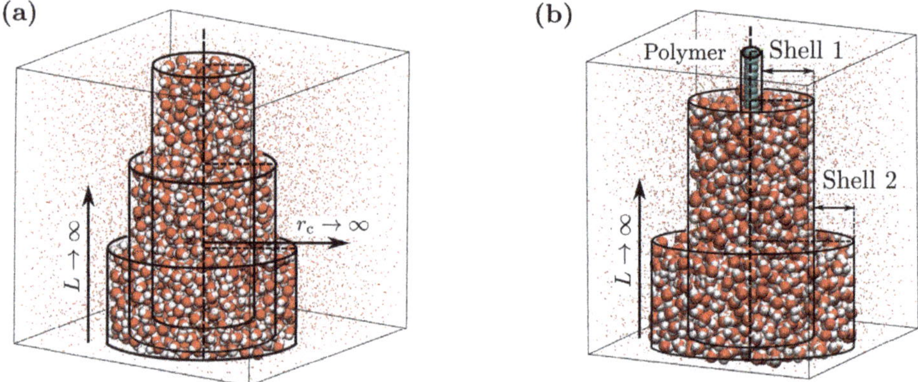

Figure 2. Observation volumes used in this work to estimate thermodynamic properties. (**a**) The isothermal compressibility of SPC/E water is estimated using cylindrical observation volumes, where the TL corresponds to both $1/L \to 0$ and $1/r_c \to 0$. (**b**) The thermodynamical quantities pertaining to polymer hydration shells are estimated using concentric cylinders as observation volume. Here, the TL corresponds to $1/L \to 0$.

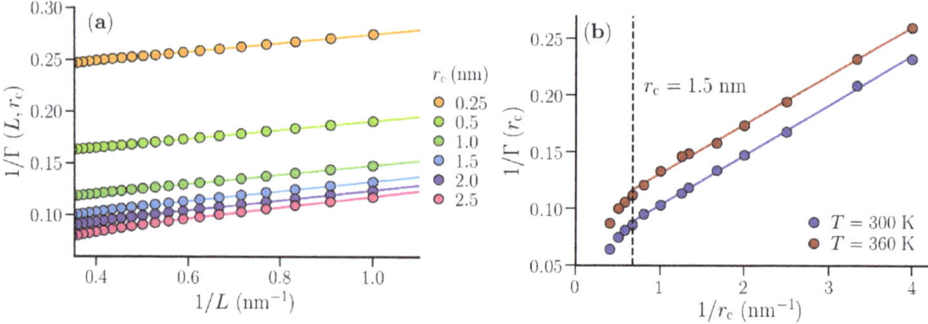

Figure 3. (**a**) $1/\Gamma(L, r_c)$ profiles for the cylindrical shells with radius r_c in SPC/E water at $T = 300$ K, as a function of $1/L$. The lines are linear fits to the data in the range $L = 1$ - 2 nm. (**b**) $1/\Gamma(r_c)$ for the cylindrical shells as a function of $1/r_c$ at $T = 300$ K, 360 K. The lines are linear fits to the data in the range $r_c < 1.5$ nm.

The data for $1/\Gamma(r_c)$ vs. $1/r_c$ presented a linear regime for $r_c < 1.5$ nm (Figure 3b). This linear regime was extrapolated to obtain $1/\Gamma^\infty$, which is the thermodynamic limiting value of $1/\Gamma$ corresponding to both $1/L$ and $1/r_c \to 0$. The values of $1/\Gamma^\infty$ (at 300 K and 360 K) were then used to estimate the bulk isothermal compressibility of SPC/E water, using Equation (2). We estimated the values of χ_T for SPC/E water to be 43.3×10^{-6} bar^{-1} and 56.0×10^{-6} bar^{-1} at 300 K and 360 K, respectively. These values are comparable to those reported by Heidari et al. [43], who used the finite size correction approach to estimate the compressibility of SPC/E water in the TL using the scaling of particle number fluctuations with system size in cubical subvolumes. The values are also in agreement with those estimated using volume fluctuation methods [44–46]: 46.1×10^{-6} bar^{-1} and 57.7×10^{-6} bar^{-1} at 298.15 K and 360 K, respectively. The agreement confirms the validity of the current implementation of the SSM. In principle, one can also calculate the compressibility for each small subvolume $\chi(L, r_c)$ and use the same $1/L$ scaling to extrapolate it to the compressibility of the infinite cylindrical shell $\chi(r_c)$, which in the limit $1/r_c \to 0$ will give the bulk isothermal compressibility χ_T [47]. Thus, with the height of the shell $L \to \infty$, one can assign $1/\Gamma(r_c)$ and $\chi(r_c)$ to be the properties of the particular shell with radius r_c. This observation asserts the rationale behind choosing cylindrical observation shells to study the thermodynamic properties of the hydration shells of a polymer.

3.2. Thermodynamic Properties of Polymer Hydration Shells

Thermodynamic properties for the individual hydration shells of the hydrophobic polymer were estimated from the fluctuations in number of water molecules therein. Unlike the cylindrical observation volumes in the SPC/E water case, the subvolumes were now concentric cylindrical shells around the linearly stretched polymer chain (Figure 2b). The boundaries of these shells were identified based on the local water number density around the polymer. At any value of α, this density was characterized by the proximal radial distribution function (pRDF), $g_p(r)$. Unlike the radial distribution function, $g(r)$, where particles are identified and binned in spherical shells, the particles in this case were binned in cylindrical shells around the polymer.

Figure 4 shows the polymer-water pRDFs for different values of the repulsive interaction strength, α. As expected, the curves shifted to larger values of r while the peak heights decreased, indicating that polymer-water density correlations decreased with increasing values of α. Correspondingly, the shell widths also increased with increasing α (Figure S1 in the Supplementary). Similar to the dependence of the effective polymer bead size, σ_p^{eff} (inset of Figure 1), on α, the pRDFs changed significantly upon increasing α between $\alpha = 1$ and $\alpha = 8$, while the changes were smaller for larger values of α. The thermodynamic properties of interest were sampled in the first two hydration shells. The inner boundary of the first hydration shell was identified as the distance r_1 at which $\rho \int_0^{r_1} 2\pi r l_z dr g_p(r) > 1$. The subsequent minima in the $g_p(r)$ curves were identified as the boundaries of the respective hydration shells.

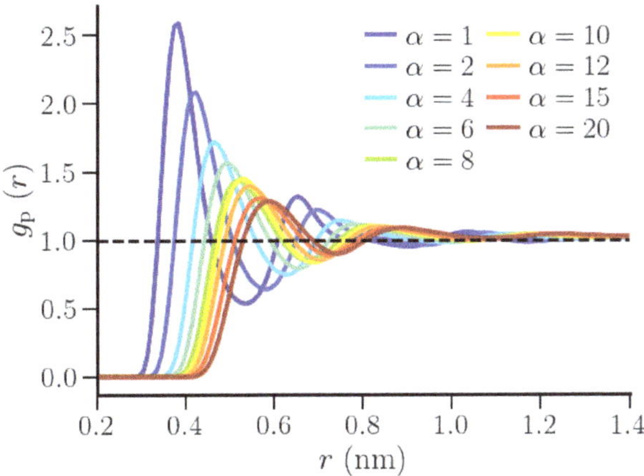

Figure 4. Proximal polymer-water RDFs for various strengths of repulsive interaction parameter α.

The particle number fluctuations were sampled and fitted to calculate the thermodynamic limiting quantities. Only the oxygen centers were used in the calculation. As in the case of SPC/E water, L was sampled in steps of 0.1 nm, starting from $L = 0.1$ nm, and the observation volumes were translated along the polymer axis (also in steps of 0.1 nm) for better statistics. Figure 5 shows the data for $1/\Gamma_s(L)$ as a function of $1/L$ for the first hydration shell of the polymer, over the range of α used in this work. The profiles were found to be linear over the intermediate L range, while the curves shifted upward with increasing α indicating enhanced density fluctuations with increasing polymer-water repulsion. The data over the range of L between 1.2 nm and 2.5 nm were fitted to a straight line to estimate $1/\Gamma_s^\infty$, which was subsequently used to estimate χ_s. The values of $1/\Gamma_s^\infty$ and χ_s estimated this way are properties of the hydration shells that are essentially infinite along the polymer axis, but finite in the other two dimensions and thus, do not pertain to a bulk system.

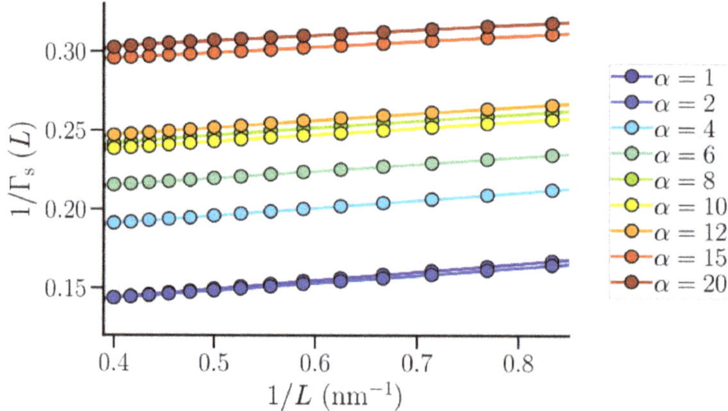

Figure 5. $1/\Gamma_s(L)$ profiles for the first polymer hydration shell for different values of the repulsive interaction parameter α. The lines are linear fits to the data in the range $L = 1.2$–2.5 nm.

To evaluate the net effect of the polymer, the same properties, here denoted as $1/\Gamma_s^{\bullet,\infty}$ and $\chi_s^{\bullet}$, were also evaluated in shells of exactly the same widths in a pure TIP4P/2005 water system where the polymer was absent. The inverse thermodynamic correction factor, $1/\Gamma_s^{\bullet,\infty}$, of these shells depended on α through the shell widths, which for the first hydration shell, increased with increasing α (Figure S1 in the Supplementary). Both $1/\Gamma_s^{\bullet,\infty}$ and $\chi_s^{\bullet}$ were found to decrease with increasing α (Figures S2a and S3a in the Supplementary). It should be noted that unlike for bulk water where the TL corresponds to both $1/L$ and $1/r_c \to 0$, the hydration shell specific thermodynamic quantities were estimated in the limit to $1/L \to 0$ alone, and therefore the values of the corresponding thermodynamic quantities are expected to be different. The respective thermodynamic quantities for polymer hydration shells and the corresponding shells in pure water were subtracted to estimate the sole effect of the polymer on the hydration shell thermodynamics for varying α,

$$\Delta 1/\Gamma_s^{\infty} \equiv 1/\Gamma_s^{\infty} - 1/\Gamma_s^{\bullet,\infty},$$
$$\Delta\chi_s \equiv \chi_s - \chi_s^{\bullet}. \tag{7}$$

Figure 6 presents $\Delta 1/\Gamma_s^{\infty}$ for the first two hydration shells of the polymer. $\Delta 1/\Gamma_s^{\infty}$ was negative for $\alpha=1$ and increased with increasing α. A similar increase in density fluctuations was observed in hydration shells of molecular sized hydrophobic solutes [24]. In the first hydration shell, $\Delta 1/\Gamma_s^{\infty}$ was negative for $\alpha < 8$, indicating that particle number fluctuations in the first hydration shell were lower than those in shells in pure water. Between $\alpha = 8$ and 12, the difference between the particle number fluctuations in the first hydration shell and the corresponding shell in pure water almost vanished. For $\alpha > 8$, the particle number fluctuations in the first hydration shell of the polymer were larger than those in the corresponding shell in pure water. The inset in Figure 6 presents $\Delta 1/\Gamma_s^{\infty}$ as a function of the effective size of the polymer beads, σ_p^{eff}. The observed dependence indicated that beyond $\sigma_p^{\mathrm{eff}} = 0.7$ nm, particle number fluctuations in the first hydration shell of the polymer suddenly increased. This observation is similar to the lengthscale crossover observed in the solvation of a hydrophobic cavity in water, when the radius of the cavity is close to 1 nm [23]. The crossover occurs as a result of the disruption of the hydrogen-bonded network of water near the surface of sufficiently large cavities. While water can readily accommodate small solutes without significantly perturbing its hydrogen-bonded network, this is no longer possible near extended surfaces of large hydrophobic solutes. As opposed to a single spherical solute, the hydrophobic polymer in our study resembled a string of small beads at low α-values, while resembling an extended hydrophobic surface when σ_p^{eff} approached 0.7 nm (keeping an equilibrium bond length of 0.153 nm).

As opposed to the first hydration shell, the density fluctuations in the second hydration shell were found to be smaller than those in the corresponding pure water shells, even at $\alpha = 20$ (Figure 6b). An increasing trend in $\Delta 1/\Gamma_s^\infty$ vs. α, similar to the trend in the first hydration shell, was observed, but with an order of magnitude difference between the two. As water molecules in the second hydration shell are not in direct contact with the polymer surface, the effect of increased polymer-water repulsion is much lower in the second hydration shell as compared to the first hydration shell. This was expected, as the fluctuations were observed to be maximum at the vicinity of the hydrophobic interface [48].

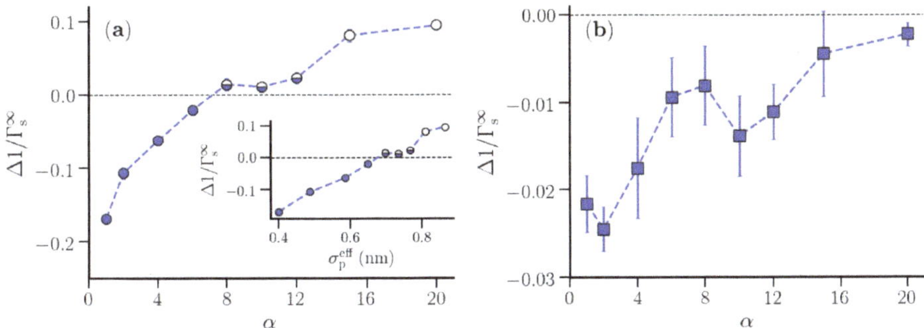

Figure 6. $\Delta 1/\Gamma_s^\infty$ profiles for the first (**a**) and second (**b**) polymer hydration shell for various strengths α of the repulsive interaction. The inset in (**a**) shows the variation of $\Delta 1/\Gamma_s^\infty$ as a function of σ_p^{eff}. In (**a**), the data points are grouped based on their variation with α (see text). The error bars are calculated over four distinct windows in the production trajectory. The lines are guide to the eyes.

While we considered the hydration shell to be infinitely long (in the limit $1/L \rightarrow 0$), realistic polymer hydration shells are strictly finite. In such a case, it is worthwhile to investigate how strongly $\Delta 1/\Gamma_s(L)$ of a finite hydration shell deviates from that in the macroscopic limit ($\Delta 1/\Gamma_s^\infty$). We expect that this difference is small for small α-values which correspond to stable hydration shells around small polymer beads with an appreciable bead-water van der Waals attraction. We here consider the variation of $1/\Gamma_s(L)$, $1/\Gamma_s^\bullet(L)$, and $\Delta 1/\Gamma_s(L)$ for finite lengths of the first hydration shell corresponding to $L = 0.4, 0.8, 2.0$, and 2.5 nm. While both $1/\Gamma_s(L)$ and $1/\Gamma_s^\bullet(L)$ were found to depend on the hydration shell height (Figure S4 in the Supplementary), $\Delta 1/\Gamma_s(L)$ was found to be independent of L in the region $\alpha \leq 8$, where it was negative (Figure 7). Interestingly, a dependence on the hydration shell dimension was observed for the systems with $\alpha > 8$, corresponding to stronger hydrophobic polymers with larger effective bead sizes and weaker bead-water van der Waals attractions. For these systems, $\Delta 1/\Gamma_s(L)$ increased with L (up to $L = 2.0$ nm), indicating that an extended hydrophobic surface created by connected polymer beads caused enhanced water density fluctuations in the first hydration shell.

Figure 8 shows the data for $\Delta \chi_s$ vs. α for the first and second hydration shell. The compressibility of both hydration shells increased with increasing α. As in the case of $\Delta 1/\Gamma_s^\infty$, the change in $\Delta \chi_s$ for the second hydration shell was rather small. For the first polymer hydration shell, the compressibility was smaller than that for the corresponding shells in pure water until $\alpha = 8$. Between $\alpha = 8$ and 12, the hydration shell compressibility was almost identical to the compressibility of the corresponding shells in pure water. For larger α-values there was a noticeable increase in the hydration shell compressibility. This is similar to the case of hydration water of hydrophobic solutes [24], where the local compressibility was found to increase for sufficiently large solutes. Figure 9 shows the tetrahedral order parameter q_{tet} of the first hydration shell of the polymer corresponding to various values of α

and those calculated for shells in pure water. For shells in pure water, q_{tet} remained almost constant around 0.66, which corresponds to the optimal hydrogen bonding capacity for TIP4P/2005 water [49]. In contrast, the tetrahedral order in the polymer hydration shell decreased by 15% upon increasing α up to $\alpha=8$, at which point $\Delta\chi_s$ became positive. The number of hydrogen bonds per water molecule in the first hydration shell showed a similar dependence (Figure S5).

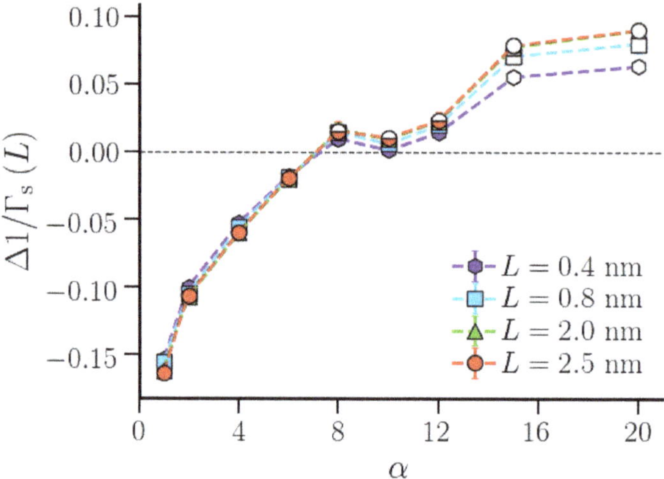

Figure 7. $\Delta 1/\Gamma_s(L)(= 1/\Gamma_s(L) - 1/\Gamma_s^\bullet(L))$ profiles for finite-sized first hydration shells with heights $L = 0.4, 0.8, 2.0$, and 2.5 nm as a function of the strength α of the repulsive interaction. The data points are grouped based on their variation with α (as in Figure 6). The error bars are calculated over four distinct windows in the production trajectory. The lines are guide to the eyes.

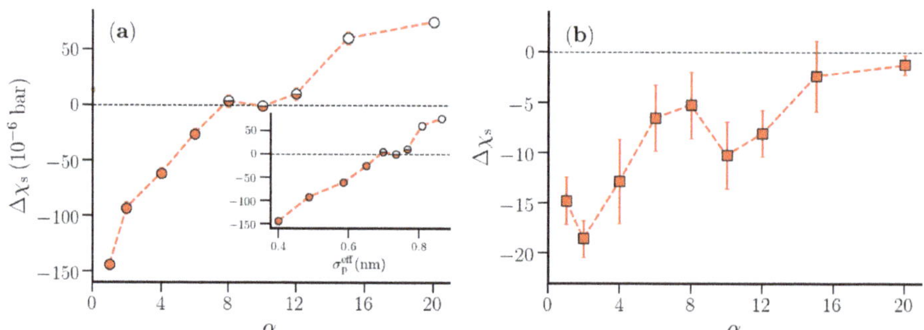

Figure 8. $\Delta\chi_s$ profiles for the first (**a**) and second (**b**) polymer hydration shell for various values of the repulsive interaction parameter α. The inset in (**a**) shows $\Delta\chi_s$ as a function of σ_p^{eff}. In (**a**), the data points are grouped based on their variation with α (see text). The error bars are calculated over four distinct windows in the production trajectory. The lines are guide to the eyes.

The increase in density fluctuations, decrease in tetrahedral order, and increase in compressibility in the first hydration shell of the hydrophobic polymer are in line with the trends expected in the case of solvation of hydrophobic solutes in water [23,24,29,50].

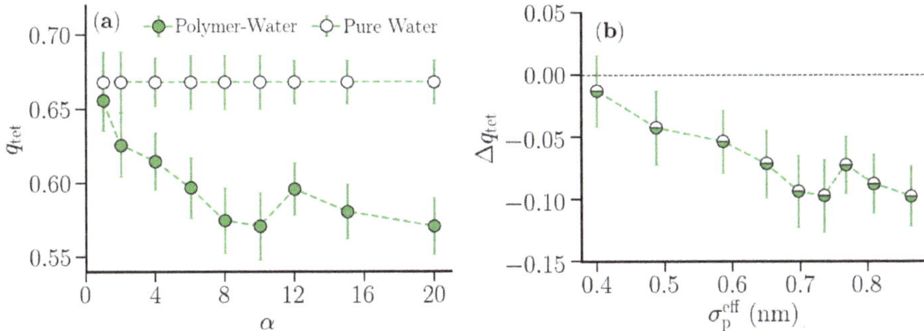

Figure 9. (a) Tetrahedral order parameter q_{tet} for the first polymer hydration shell and the corresponding shell in pure water for various strengths α of the repulsive interaction. (b) $\Delta q_{tet} = q_{tet}(\text{polymer shell}) - q_{tet}(\text{water shell})$ as a function of σ_p^{eff}. The error bars are calculated over four distinct windows in the production trajectory. The lines are guide to the eyes.

4. Scope and Limitations

The polymer analyzed in this work is a generic homo-polymer in a restrained, stretched conformation, where the monomeric units are represented by simple beads without any side chains. It is essentially the simplest possible polymer system that could be analyzed in the current framework. Applying the current approach to study chemically realistic polymer systems requires that several technical challenges are addressed. Hill's formalism [12,13] is framed for a large ensemble of equivalent, independent, and distinguishable small systems. In the present context, this implies that the small system must be at least the size of a monomeric unit and, therefore, L has to be chosen as an integer multiple of the monomer length. Therefore, to effectively sample the density fluctuations, very long chemically realistic polymer chains (much longer than the one considered here) need to be considered in the simulations. For the generic polymer used in this work, $1/\Gamma$ is insensitive to whether density fluctuations within the small system L are sampled by shifting it along the polymer backbone in steps of 0.1 nm or in steps that correspond to a monomeric unit, essentially due to its linearly stretched configuration. The same will not hold for flexible polymers as the monomeric units themselves will differ in terms of the volume they occupy and, therefore, static observation volumes can no longer be used to sample the fluctuations. Therefore, additional methodological progress requires implementation of dynamical observation volumes.

In addition to the application of the SSM to a generic polymer model in pure water, mixed solvents may be considered to be well. In this context, the SSM can be used to provide information on excess chemical potentials and partial molar enthalpies and entropies of solvent components in the solvation shell of the polymer based on analyses of energy and particle number fluctuations and the calculation of KBIs. In particular, in systems where cosolvents and small organic molecules preferentially bind to the polymer, a full characterization of solvation shell properties is needed to address open questions related to the corresponding changes in aqueous polymer solubility [51–53]. The SSM provides a route to calculate these properties.

5. Conclusions

We have extended the SSM to study the hydration shell compressibility of a generic hydrophobic polymer in water. We identified the hydration shells based on the proximal distribution of water molecules around a linearly extended polymer chain. Water density fluctuations were sampled in small concentric cylindrical shells around the chain, which in the limit of the shell height $L \to \infty$ resulted in thermodynamic quantities that could be assigned to the hydration shells. We systematically varied the range of the polymer-water repulsion and observed a crossover behavior in the excess inverse thermodynamic correction factor ($\Delta 1/\Gamma_s^\infty$) and the excess compressibility ($\Delta\chi_s$) of the hydration shells, defined relative to the properties of the same shells in pure water (without polymer). The negative-to-positive crossover observed in $\Delta 1/\Gamma_s^\infty$ and in $\Delta\chi_s$ happened at an intermediate range of the polymer-water repulsion, where the effective polymer bead diameter was around 0.7 nm. We also observed a complementary trend in the tetrahedral order parameter of the shell, which measured the deviation in the hydrogen bonding coordination of water molecules from the ideal tetrahedrally hydrogen-bonded structure.

The observations made in this work were in line with those reported for solvation of spherical hydrophobic solutes in water [23,24,50]. It is well established that small hydrophobic cavities/ solutes are solvated by restructuring of water hydrogen bonds around them, while solvation of large hydrophobic cavities occurs via breakage of water hydrogen bonds near the solute surface. The signatures of these two distinct solvation mechanisms were observed in the present work in the excess polymer hydration shell compressibility.

Supplementary Materials: The following are available online at http://www.mdpi.com/2079-4991/10/8/1460/ s1, Figure S1: Width of polymer hydration shells vs. α, Figure S2: $1/\Gamma_s^\infty$ and $1/\Gamma_s^{\bullet,\infty}$ vs. α , Figure S3: χ_s and $\chi_s^\bullet$ vs. α, Figure S4: $1/\Gamma_s(L)$ and $1/\Gamma_s^\bullet(L)$ vs. α, Figure S5:$\Delta n_{H-bonds} = n_{H-bonds} - n_{H-bonds}^\bullet$ vs. α.

Author Contributions: Conceptualization, N.F.A.v.d.V.; methodology, M.T., S.B., S.J.B.; formal analysis, M.T., S.B., S.J.B. and N.F.A.v.d.V.; writing–original draft preparation, M.T.; writing–review and editing, S.B., S.J.B. and N.F.A.v.d.V. All authors have read and agreed to the published version of the manuscript.

Funding: This research was funded by the Deutsche Forschungsgemeinschaft (DFG) through the Collaborative Research Center Transregio TRR 146 Multiscale Simulation Methods for Soft Matter Systems.

Acknowledgments: Computations for this work were performed on the Lichtenberg High Performance Computer of Technische Universität Darmstadt, Germany.

Conflicts of Interest: The authors declare no conflict of interest.

Abbreviations

The following abbreviations are used in this manuscript:

TL	Thermodynamic Limit
SSM	Small-System Method
KBI	Kirkwood-Buff Integral
RDF	Radial Distribution Function
pRDF	Proximal pair correlation function

References

1. Frenkel, D.; Smit, B. *Understanding Molecular Simulation: From Algorithms to Applications*, 2nd ed.; Academic Press, Inc.: New York, NY, USA, 2001.
2. Allen, M.P.; Tildesley, D.J. *Computer Simulation of Liquids*, 2nd ed.; Clarendon Press: Oxford, UK, 2017. [CrossRef]
3. Lebowitz, J.L.; Percus, J.K. Long-Range Correlations in a Closed System with Applications to Nonuniform Fluids. *Phys. Rev.* **1961**, *122*, 1675–1691. [CrossRef]
4. Perera, A.; Zoranić, L.; Sokolić, F.; Mazighi, R. A Comparative Molecular Dynamics Study of Water-Methanol and Acetone-Methanol Mixtures. *J. Mol. Liq.* **2011**, *159*, 52–59. [CrossRef]
5. Ganguly, P.; van der Vegt, N.F.A. Convergence of Sampling Kirkwood-Buff Integrals of Aqueous Solutions with Molecular Dynamics Simulations. *J. Chem. Theory Comput.* **2013**, *9*, 1347–1355. [CrossRef] [PubMed]
6. Krüger, P.; Schnell, S.K.; Bedeaux, D.; Kjelstrup, S.; Vlugt, T.J.H.; Simon, J.M. Kirkwood-Buff Integrals for Finite Volumes. *J. Phys. Chem. Lett.* **2013**, *4*, 235–238. [CrossRef] [PubMed]
7. Krüger, P.; Vlugt, T.J.H. Size and Shape Dependence of Finite-Volume Kirkwood-Buff Integrals. *Phys. Rev. E* **2018**, *97*, 051301. [CrossRef]
8. Dawass, N.; Krüger, P.; Schnell, S.K.; Bedeaux, D.; Kjelstrup, S.; Simon, J.M.; Vlugt, T.J.H. Finite-Size Effects of Kirkwood-Buff Integrals from Molecular Simulations. *Mol. Simul.* **2018**, *44*, 599–612. [CrossRef]

9.   Milzetti, J.; Nayar, D.; van der Vegt, N.F.A. Convergence of Kirkwood–Buff Integrals of Ideal and Nonideal Aqueous Solutions Using Molecular Dynamics Simulations. *J. Phys. Chem. B* **2018**, *122*, 5515–5526. [CrossRef]

10. Schnell, S.K.; Vlugt, T.J.H.; Simon, J.M.; Bedeaux, D.; Kjelstrup, S. Thermodynamics of a Small System in a μT Reservoir. *Chem. Phys. Lett.* **2011**, *504*, 199–201. [CrossRef]

11. Schnell, S.K.; Liu, X.; Simon, J.M.; Bardow, A.; Bedeaux, D.; Vlugt, T.J.H.; Kjelstrup, S. Calculating Thermodynamic Properties from Fluctuations at Small Scales. *J. Phys. Chem. B* **2011**, *115*, 10911–10918. [CrossRef]

12. Hill, T.L. Thermodynamics of Small Systems. *J. Chem. Phys.* **1962**, *36*, 3182–3197. [CrossRef]

13. Hill, T.L. *Thermodynamics of Small Systems*; Dover Publications: Mineola, NY, USA, 1994.

14. Schnell, S.K.; Skorpa, R.; Bedeaux, D.; Kjelstrup, S.; Vlugt, T.J.H.; Simon, J.M. Partial Molar Enthalpies and Reaction Enthalpies from Equilibrium Molecular Dynamics Simulation. *J. Chem. Phys.* **2014**, *141*, 144501. [CrossRef] [PubMed]

15. Strøm, B.A.; Simon, J.M.; Schnell, S.K.; Kjelstrup, S.; He, J.; Bedeaux, D. Size and Shape Effects on the Thermodynamic Properties of Nanoscale Volumes of Water. *Phys. Chem. Chem. Phys.* **2017**, *19*, 9016–9027. [CrossRef] [PubMed]

16. Schnell, S.K.; Englebienne, P.; Simon, J.M.; Kruger, P.; Balaji, S.P.; Kjelstrup, S.; Bedeaux, D.; Bardow, A.; Vlugt, T.J.H. How to Apply the Kirkwood-Buff Theory to Individual Species in Salt Solutions. *Chem. Phys. Lett.* **2013**, *582*, 154–157. [CrossRef]

17. Erdös, M.; Galteland, O.; Bedeaux, D.; Kjelstrup, S.; Moultos, O.; Vlugt, T. Gibbs Ensemble Monte Carlo Simulation of Fluids in Confinement: Relation between the Differential and Integral Pressures. *Nanomaterials* **2020**, *10*, 293. [CrossRef] [PubMed]

18. Rauter, M.; Galteland, O.; Erdös, M.; Moultos, O.; Vlugt, T.; Schnell, S.; Bedeaux, D.; Kjelstrup, S. Two-Phase Equilibrium Conditions in Nanopores. *Nanomaterials* **2020**, *10*, 608. [CrossRef]

19. Jamali, S.H.; Bardow, A.; Vlugt, T.J.H.; Moultos, O.A. Generalized Form for Finite-Size Corrections in Mutual Diffusion Coefficients of Multicomponent Mixtures Obtained from Equilibrium Molecular Dynamics Simulation. *J. Chem. Theory Comput.* **2020**, *16*, 3799–3806. [CrossRef]

20. Trinh, T.T.; Bedeaux, D.; Simon, J.M.; Kjelstrup, S. Thermodynamic Characterization of Two Layers of CO_2 on a Graphite Surface. *Chem. Phys. Lett.* **2014**, *612*, 214–218. [CrossRef]

21. Trinh, T.T.; van Erp, T.S.; Bedeaux, D.; Kjelstrup, S.; Grande, C.A. A Procedure to Find Thermodynamic Equilibrium Constants for CO_2 and CH_4 Adsorption on Activated Carbon. *Phys. Chem. Chem. Phys.* **2015**, *17*, 8223–8230. [CrossRef]

22. Lum, K.; Chandler, D.; Weeks, J.D. Hydrophobicity at Small and Large Length Scales. *J. Phys. Chem. B* **1999**, *103*, 4570–4577. [CrossRef]

23. Chandler, D. Interfaces and the Driving Force of Hydrophobic Assembly. *Nature* **2005**, *437*, 640–647. [CrossRef]

24. Sarupria, S.; Garde, S. Quantifying Water Density Fluctuations and Compressibility of Hydration Shells of Hydrophobic Solutes and Proteins. *Phys. Rev. Lett.* **2009**, *103*, 037803. [CrossRef] [PubMed]

25. Mittal, J.; Hummer, G. Static and Dynamic Correlations in Water at Hydrophobic Interfaces. *Proc. Natl. Acad. Sci. USA* **2008**, *105*, 20130–20135. [CrossRef] [PubMed]

26. Huang, D.M.; Geissler, P.L.; Chandler, D. Scaling of Hydrophobic Solvation Free Energies. *J. Phys. Chem. B* **2001**, *105*, 6704–6709. [CrossRef]

27. Patel, A.J.; Varilly, P.; Chandler, D. Fluctuations of Water Near Extended Hydrophobic and Hydrophilic Surfaces. *J. Phys. Chem. B* **2010**, *114*, 1632–1637. [CrossRef]

28. Rego, N.B.; Xi, E.; Patel, A.J. Protein Hydration Waters Are Susceptible to Unfavorable Perturbations. *J. Am. Chem. Soc.* **2019**, *141*, 2080–2086. [CrossRef]

29. Rajamani, S.; Truskett, T.M.; Garde, S. Hydrophobic Hydration from Small to Large Lengthscales: Understanding and Manipulating the Crossover. *Proc. Natl. Acad. Sci. USA* **2005**, *102*, 9475–9480. [CrossRef]

30. Zangi, R.; Zhou, R.; Berne, B.J. Urea's Action on Hydrophobic interactions. *J. Am. Chem. Soc.* **2009**, *131*, 1535–1541. [CrossRef]

31. Kanduč, M.; Chudoba, R.; Palczynski, K.; Kim, W.K.; Roa, R.; Dzubiella, J. Selective Solute Adsorption and Partitioning Around Single PNIPAM Chains. *Phys. Chem. Chem. Phys.* **2017**, *19*, 5906–5916. [CrossRef]

32. Abascal, J.L.F.; Vega, C. A General Purpose Model for the Condensed Phases of Water: TIP4P/2005. *J. Chem. Phys.* **2005**, *123*, 234505. [CrossRef]

33. Abraham, M.J.; Murtola, T.; Schulz, R.; Páll, S.; Smith, J.C.; Hess, B.; Lindahl, E. GROMACS: High Performance Molecular Simulations Through Multi-Level Parallelism from Laptops to Supercomputers. *SoftwareX* **2015**, *1-2*, 19–25. [CrossRef]

34. Hess, B. P-LINCS: A Parallel Linear Constraint Solver for Molecular Simulation. *J. Chem. Theory Comput.* **2008**, *4*, 116–122. [CrossRef] [PubMed]

35. Darden, T.; York, D.; Pedersen, L. Particle Mesh Ewald: An $N \cdot \log N$ Method for Ewald Sums in Large Systems. *J. Chem. Phys.* **1993**, *98*, 10089–10092. [CrossRef]

36. Bussi, G.; Donadio, D.; Parrinello, M. Canonical Sampling Through Velocity Rescaling. *J. Chem. Phys.* **2007**, *126*, 014101. [CrossRef]

37. Nosé, S. A Molecular Dynamics Method for Simulations in the Canonical Ensemble. *Mol. Phys.* **1984**, *52*, 255–268. [CrossRef]

38. Berendsen, H.J.C.; Postma, J.P.M.; van Gunsteren, W.F.; DiNola, A.; Haak, J.R. Molecular Dynamics with Coupling to an External Bath. *J. Chem. Phys.* **1984**, *81*, 3684–3690. [CrossRef]

39. Parrinello, M.; Rahman, A. Polymorphic Transitions in Single Crystals: A New Molecular Dynamics Method. *J. Appl. Phys.* **1981**, *52*, 7182–7190. [CrossRef]

40. Berendsen, H.J.C.; Grigera, J.R.; Straatsma, T.P. The Missing Term in Effective Pair Potentials. *J. Phys. Chem.* **1987**, *91*, 6269–6271. [CrossRef]

41. Chau, P.L.; Hardwick, A.J. A New Order Parameter for Tetrahedral Configurations. *Mol. Phys.* **1998**, *93*, 511–518. [CrossRef]

42. Errington, J.R.; Debenedetti, P.G. Relationship Between Structural Order and the Anomalies of Liquid Water. *Nature* **2005**, *409*, 318–321. [CrossRef]

43. Heidari, M.; Kremer, K.; Potestio, R.; Cortes-Huerto, R. Finite-Size Integral Equations in the Theory of Liquids and the Thermodynamic Limit in Computer Simulations. *Mol. Phys.* **2018**, *116*, 3301–3310. [CrossRef]

44. Pi, H.L.; Aragones, J.L.; Vega, C.; Noya, E.G.; Abascal, J.L.; Gonzalez, M.A.; McBride, C. Anomalies in Water as Obtained from Computer Simulations of the TIP4P/2005 Model: Density Maxima, and Density, Isothermal Compressibility and Heat Capacity Minima. *Mol. Phys.* **2009**, *107*, 365–374. [CrossRef]

45. Pathak, H.; Palmer, J.C.; Schlesinger, D.; Wikfeldt, K.T.; Sellberg, J.A.; Pettersson, L.G.M.; Nilsson, A. The Structural Validity of Various Thermodynamical Models of Supercooled Water. *J. Chem. Phys.* **2016**, *145*, 134507. [CrossRef] [PubMed]

46. Pettersson, L.G.; Nilsson, A. The Structure of Water; From Ambient to Deeply Supercooled. *J. Non-Cryst. Solids* **2015**, *407*, 399–417. [CrossRef]

47. Román, F.L.; White, J.A.; Velasco, S. Fluctuations in an Equilibrium Hard-Disk Fluid: Explicit Size Effects. *J. Chem. Phys.* **1997**, *107*, 4635–4641. [CrossRef]

48. Jamadagni, S.N.; Godawat, R.; Garde, S. Hydrophobicity of Proteins and Interfaces: Insights from Density Fluctuations. *Annu. Rev. Chem. Biomol. Eng.* **2011**, *2*, 147–171. [CrossRef] [PubMed]

49. Duboué-Dijon, E.; Laage, D. Characterization of the Local Structure in Liquid Water by Various Order Parameters. *J. Phys. Chem. B* **2015**, *119*, 8406–8418. [CrossRef]

50. Stillinger, F.H. Structure in Aqueous Solutions of Nonpolar Solutes from the Standpoint of Scaled-Particle Theory. *J. Solut. Chem.* **1973**, *2*, 141–158. [CrossRef]

51. Nayar, D.; van der Vegt, N.F.A. Cosolvent Effects on Polymer Hydration Drive Hydrophobic Collapse. *J. Phys. Chem. B* **2018**, *122*, 3587–3595. [CrossRef]

52. Dalgicdir, C.; Rodríguez-Ropero, F.; van der Vegt, N.F.A. Computational Calorimetry of PNIPAM Cononsolvency in Water/Methanol Mixtures. *J. Phys. Chem. B* **2017**, *121*, 7741–7748. [CrossRef]

53. Rodríguez-Ropero, F.; van der Vegt, N.F.A. On the Urea Induced Hydrophobic Collapse of a Water Soluble Polymer. *Phys. Chem. Chem. Phys.* **2015**, *17*, 8491–8498. [CrossRef]

Article

Gibbs Ensemble Monte Carlo Simulation of Fluids in Confinement: Relation between the Differential and Integral Pressures

Máté Erdős [1], Olav Galteland [2], Dick Bedeaux [2], Signe Kjelstrup [2], Othonas A. Moultos [1] and Thijs J. H. Vlugt [1,*]

[1] Engineering Thermodynamics, Process & Energy Department, Faculty of Mechanical, Maritime and Materials Engineering, Delft University of Technology, Leeghwaterstraat 39, 2628CB Delft, The Netherlands; m.erdos-2@tudelft.nl (M.E.); o.moultos@tudelft.nl (O.A.M.)

[2] PoreLab, Department of Chemistry, Norwegian University of Science and Technology, 7031 Trondheim, Norway; olav.galteland@ntnu.no (O.G.); dick.bedeaux@ntnu.no (D.B.); signe.kjelstrup@ntnu.no (S.K.)

* Correspondence: t.j.h.vlugt@tudelft.nl

Received: 12 January 2020; Accepted: 5 February 2020; Published: 9 February 2020

Abstract: The accurate description of the behavior of fluids in nanoporous materials is of great importance for numerous industrial applications. Recently, a new approach was reported to calculate the pressure of nanoconfined fluids. In this approach, two different pressures are defined to take into account the smallness of the system: the so-called differential and the integral pressures. Here, the effect of several factors contributing to the confinement of fluids in nanopores are investigated using the definitions of the differential and integral pressures. Monte Carlo (MC) simulations are performed in a variation of the Gibbs ensemble to study the effect of the pore geometry, fluid-wall interactions, and differential pressure of the bulk fluid phase. It is shown that the differential and integral pressure are different for small pores and become equal as the pore size increases. The ratio of the driving forces for mass transport in the bulk and in the confined fluid is also studied. It is found that, for small pore sizes (i.e., $<5\sigma_{\text{fluid}}$), the ratio of the two driving forces considerably deviates from 1.

Keywords: nanothermodynamics; porous systems; molecular simulation; differential pressure; integral pressure

---

1. Introduction

The widespread application of nanoporous materials in several fields, such as chromatography, membrane separation, catalysis, etc., has lead to a growing interest in the accurate description of the thermodynamic behavior of fluids confined in nanopores [1–6]. The pressure of a nanoconfined fluid is one of the most important thermodynamic properties which is needed for an accurate description of the flow rate, diffusion coefficient, and the swelling of the nanoporous material [7–10]. Various approaches for calculating the pressure of a fluid in a nanopore have been proposed [1,11–13]. The main difficulty of the pressure calculation arises from the ambiguous definition of the pressure tensor inside porous materials due to the presence of curved surfaces and confinement effects [7,14,15]. Traditional thermodynamic laws and concepts, such as Gibbs surface dynamics, Kelvin equation, etc., may not be applicable at the nano-scale [16]. In the past decade, several methods were reported using different simulation techniques i.e., classical density functional theory [17,18], equation of state modeling [19], etc., to model the behavior of fluids in confinement.

Recently, Galteland et al. [11] reported a new approach for the calculation of pressure in nanoporous materials using Hill's thermodynamics for small systems [20]. In this approach, two different pressures are needed to account for confinement effects in nanoporous materials: the differential pressure P, and the integral pressure $\hat{p}$. In an ensemble of small systems, the differential pressure times the volume change of the small systems is equal to the work exerted on the surroundings by the volume change. The differential pressure corresponds to the macroscopic pressure and does not depend on the size of the system. The addition of a small system to an ensemble of small systems exerts work on the surroundings which is equal to the integral pressure times the volume of the added small system. The differential and integral pressures are different for small systems and become equal in the thermodynamic limit. For a system with volume V, the two pressures are related by [11,20]:

$$P(V) = \left(\frac{\partial(\hat{p}(V)V)}{\partial V} \right)_{T,\mu} = \hat{p}(V) + V \left(\frac{\partial(\hat{p}(V))}{\partial V} \right)_{T,\mu}. \tag{1}$$

As shown in Equation (1), the two pressures are different only when the integral pressure $\hat{p}$ depends on the volume of the system. Galteland et al. [11] performed equilibrium and non-equilibrium molecular dynamics simulations of Lennard–Jones (LJ) fluids in a face-centered lattice of spherical grains representing a porous medium. Using Hill's thermodynamics of small systems [20] and the additive property of the grand potential, Galteland et al. [11] defines the compressional energy ($\hat{p}V$) of the representative elementary volume (REV) in the simulations as follows:

$$\hat{p}V = \hat{p}^{f}V^{f} + \hat{p}^{r}V^{r} - \hat{\gamma}^{fr}\omega^{fr}, \tag{2}$$

where $\hat{p}$ is the integral pressure of volume V, $\hat{p}^{f}$ and V^{f} are the integral pressure and volume of the fluid, $\hat{p}^{r}$ and V^{r} are the integral pressure and volume of the grain particles, and the $\hat{\gamma}^{fr}$ and ω^{fr} are the integral surface tension and surface area between the fluid and grain particles. Based on the obtained results, it was concluded that the definition of two pressures is needed to calculate the pressure of the fluid in nanoporous medium.

In this study, the relation between the differential and integral pressure is investigated by performing Monte Carlo (MC) simulations of LJ fluids. The simulations are carried out in a modified Gibbs ensemble, using two simulation boxes in equilibrium with each other. One box represents the bulk fluid, while the other simulation box represents the nanoconfined system, including walls that interact with the fluid particles, as shown in Figure 1. The effect of confinement on the integral pressure is investigated by considering different fluid-wall interaction strengths and pore geometries, namely, a cylinder and a slit pore. To investigate the relation between the differential and integral pressure, the difference of the two pressures, $P - \langle \hat{p} \rangle$, and the ratio of driving forces for mass transport, $\dfrac{\mathrm{d}\langle \hat{p} \rangle}{\mathrm{d}P}$, are computed. In Section 2, the devised ensemble and the equations used to calculate the pressure and energy of the system are presented. In Section 3, a rigorous derivation of the used expression to compute the ratio of driving forces is shown. In Section 4, the results for the different differential pressures and pore geometries are shown. In Section 5, our conclusions are summarized.

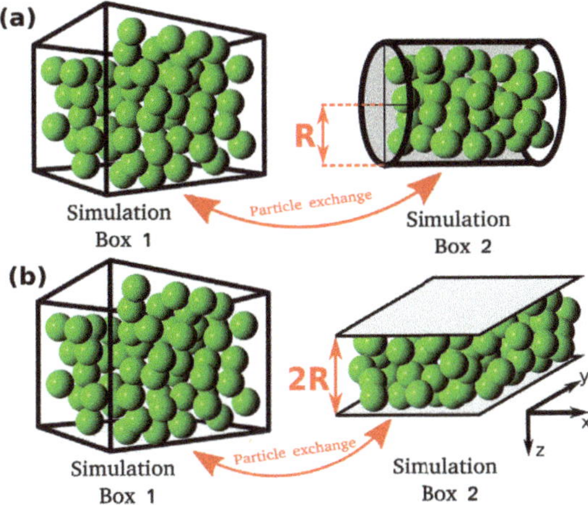

Figure 1. Schematic representation of the Monte Carlo (MC) simulation scheme. Simulation Box 1 represents the bulk fluid with differential pressure P. Simulation Box 2 contains the confined fluid with (average) integral pressure $\langle \hat{p} \rangle$. (**a,b**) The two investigated systems are shown where the bulk fluid is in equilibrium with the nanoconfined fluid in a cylinder and in a slit pore, respectively. Due to the particle exchange, the chemical potential of the two boxes are equal, but in general $P \neq \langle \hat{p} \rangle$.

2. Simulation Details

All MC simulations are carried out using an in-house simulation code. The MC simulations consist of two simulation domains, Simulation Box 1 and Simulation box 2 (see Figure 1). Throughout the manuscript the terms Simulation Box 1 and 2 are used to refer to the two domains, however, simulation domain 2 does not correspond to an actual box. The total number of particles in the system, N_T, is fixed and particles can be exchanged between the simulation boxes. Box 1 is cubic and has periodic boundary conditions imposed in all directions. Box 1 represents a bulk fluid. The differential pressure, P, and temperature, T, in Simulation Box 1 are imposed, while the volume, V_1, and the number of particles, N_1, can fluctuate. Simulation Box 2 is a cylinder or a slit pore with a fixed volume V_2. The size of the cylinder and the slit pore are defined by the radius, R, and by the distance between the two parallel planes, $2R$, respectively (Figure 1). In Box 2, periodic boundary conditions are applied only in the axial direction for the cylinder and in the x and y directions for the slit pore (see Figure 1). Box 2 represents the confined fluid which has an integral pressure $\hat{p}$. In Simulation Box 2, the volume, V_2, and temperature T are imposed, while the number of particles N_2 can fluctuate by exchanging particles with Box 1. The instantaneous integral pressure fluctuates, and by definition [20] its ensemble average, $\langle \hat{p} \rangle$, will be equal to P only for macroscopic systems ($R \rightarrow \infty$). The ensemble used is a variation of the NPT-Gibbs ensemble [21]. The main difference between the ensemble used in this work and the conventional NPT-Gibbs ensemble is that in our simulations the volume of Box 2 is fixed [21]. Essentially, Box 2 corresponds to the grand canonical ensemble with the reservoir explicitly modeled in Box 1. This computational setup was also used in other studies, i.e., A. Z. Panagiotopoulos et al. [21], P. Bai et al. [22], etc.

In the MC simulations, three types of trial moves are used: translation, volume change, and particle exchange. The translation and particle exchange trial moves are used in both simulation boxes, while volume change trial moves are only performed in Simulation Box 1. The acceptance rules of the trial moves can be found elsewhere [23]. The particle exchange trial move, ensures that Box 1 and Box 2 are in chemical equilibrium, i.e., the chemical potentials of the two boxes are equal ($\mu_1 = \mu_2$). The chemical potentials of the boxes are defined as the sum of the ideal and excess chemical potentials of the fluid in the respective box ($\mu_1 = \mu_1^{\text{id}} + \mu_1^{\text{ex}}$, $\mu_2 = \mu_2^{\text{id}} + \mu_2^{\text{ex}}$). The ideal gas chemical potentials ($\mu_1^{\text{id}}, \mu_2^{\text{id}}$) are calculated based on the density and temperature of the fluid. The excess chemical potential

$(\mu_1^{\text{ex}}, \mu_2^{\text{ex}})$ can be calculated using different methods, i.e., Widom's test particle insertion method [23], Contionous Fractional Component Monte Carlo method [24,25], Bennet acceptance ratio method [26], etc. Although the Bennet acceptance ratio method is computationally more efficient than the Widom's test particle insertion method, the Widom's test particle method is sufficient for the systems considered in this study. Separate simulations are carried out using different cylinder radii R, while imposing $P = 0.2$ (in reduced units) in Box 1. In Figure 2, the chemical potential of Box 1 and Box 2 (Figure 2a), as well as the density of Box 1 and Box 2 (Figure 2b), are shown as a function of the cylinder radius at $P = 0.2$. As can be seen from Figure 2a, the chemical potentials of Box 1 and Box 2 are equal within the uncertainties of the simulations.

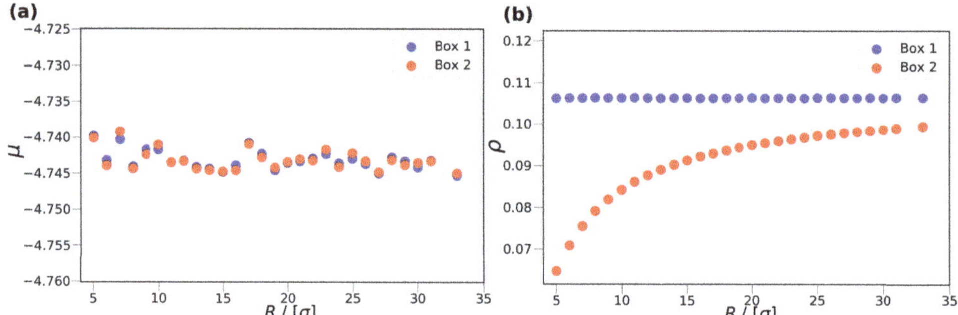

Figure 2. The chemical potential (**a**) and density (**b**) of the bulk, Box 1, and confined fluids, Box 2, as a function of the cylinder radius, R, at $P = 0.2$ (in Simulation Box 1). The blue and red colors represent Box 1 and Box 2, respectively. The temperature is fixed at $T = 2$. All values are presented in reduced units. The error bars are smaller than the symbol sizes.

In all simulations, the total potential energy, U, is calculated using the 12-6 LJ interaction potential:

$$U = U_{\text{fluid}-\text{fluid}} + U_{\text{fluid}-\text{wall}},\tag{3}$$

where $U_{\text{fluid}-\text{fluid}}$ is the potential energy due to interaction between the fluid particles, and $U_{\text{fluid}-\text{wall}}$ represents the potential energy contribution from the interactions between the fluid particles and the wall of Box 2. $U_{\text{fluid}-\text{fluid}}$ in both simulation boxes is calculated according to:

$$U_{\text{fluid}-\text{fluid}} = \begin{cases} \sum_{i<j} \left[4\epsilon_{\text{fluid}}\left(\left(\frac{\sigma_{\text{fluid}}}{r_{ij}}\right)^{12} - \left(\frac{\sigma_{\text{fluid}}}{r_{ij}}\right)^{6} \right) - U_{\text{shift}} \right] & r_{ij} < 2.5\sigma_{\text{fluid}} \\ 0, & \text{otherwise} \end{cases},\tag{4}$$

where r_{ij} is the distance of particle i and j, U_{shift} makes the interaction potentially continuous at the cut-off distance ($2.5\sigma_{\text{fluid}}$), and ϵ_{fluid}, σ_{fluid} are the LJ parameters. In the past, several studies were reported using different types of interaction potentials to model the fluid-solid interactions in confined spaces [22,27,28]. Since the aim of this study is to show the difference between the differential and integral pressures and not to simulate some specific adsorption system, only two types of interaction potentials are considered for the interaction of fluid particles with the wall in Simulation Box 2. In the first case, the wall has only repulsive interactions with the fluid particles. The potential energy contribution of this type is calculated based on the Weeks–Chandler–Andersen potential [29]:

$$U_{\text{fluid}-\text{wall}} = \begin{cases} \sum_{i=0}^{N_2} \left[4\epsilon_{\text{fw}}\left(\left(\frac{\sigma_{\text{fw}}}{r_{wi}}\right)^{12} - \left(\frac{\sigma_{\text{fw}}}{r_{wi}}\right)^{6} \right) + 1 \right] & r_{wi} < 2^{1/6}\sigma_{\text{fw}} \\ 0, & \text{otherwise} \end{cases},\tag{5}$$

where N_2 is the number of particles in Box 2, r_{wi} is the closest distance of particle i from the walls, and ϵ_{fw} and σ_{fw} are the LJ parameters for the interaction between the wall and the fluid particles. In the second case, the attractive interactions between the wall and fluid particles are taken into account using the traditional form of the 12-6 LJ interaction potential:

$$U_{\text{fluid}-\text{wall}} = \begin{cases} \sum_{i=0}^{N_2} \left[4\epsilon_{fw} \left(\left(\frac{\sigma_{fw}}{r_{wi}} \right)^{12} - \left(\frac{\sigma_{fw}}{r_{wi}} \right)^6 \right) - U_{\text{shift}-fw} \right] & r_{wi} < 2.5\sigma_{fw} \\ 0, & \text{otherwise} \end{cases} \tag{6}$$

where $U_{\text{shift}-fw}$ makes the interaction potential continuous at the cut-off distance, and ϵ_{fw} and σ_{fw} are the LJ parameters.

The expression for calculating the integral pressure in Box 2 is as follows[30]:

$$\hat{p} = \frac{N_2}{V_2} k_B T + \hat{p}_{\text{fluid}-\text{fluid}} + \hat{p}_{\text{fluid}-\text{wall}}, \tag{7}$$

where $\hat{p}_{\text{fluid}-\text{fluid}}$ represents the contribution of the fluid-fluid interaction to the integral pressure, k_B is the Boltzmann constant, and $\hat{p}_{\text{fluid}-\text{wall}}$ represents the contribution of the fluid-wall interaction to the integral pressure. The $\hat{p}_{\text{fluid}-\text{fluid}}$ and $\hat{p}_{\text{fluid}-\text{wall}}$ terms represent the virial contributions of the integral pressure. The terms are calculated based on the virial theorem [7], i.e., using the derivative of the potential energy function with respect to r. The pressure term $\hat{p}_{\text{fluid}-\text{fluid}}$ is calculated as follows [7]:

$$\hat{p}_{\text{fluid}-\text{fluid}} = \frac{48}{3V_2} \epsilon_{\text{fluid}} \sum_{i<j} \left(\left(\frac{\sigma_{\text{fluid}}}{r_{ij}} \right)^{12} - 0.5 \left(\frac{\sigma_{\text{fluid}}}{r_{ij}} \right)^6 \right). \tag{8}$$

The third term in Equation (7) represents the pressure contribution related to the interactions of the LJ particles with the wall of Box 2. The term $\hat{p}_{\text{fluid}-\text{wall}}$ for the repulsive wall potential is derived based on Equation (5) [7]. The following expression is obtained:

$$\hat{p}_{\text{fluid}-\text{wall}} = \begin{cases} \sum_{i=1}^{N_2} \frac{24}{3V_2} \epsilon_{fw} \left(\left(\frac{\sigma_{fw}}{r_{wi}} \right)^{12} - 0.5 \left(\frac{\sigma_{fw}}{r_{wi}} \right)^6 \right) & r_{wi} < 2^{1/6}\sigma_{fw} \\ 0, & \text{otherwise} \end{cases} \tag{9}$$

The following expression is used to calculate the $\hat{p}_{\text{fluid}-\text{wall}}$ when the wall has also attractive interactions with the fluid [7]:

$$\hat{p}_{\text{fluid}-\text{wall}} = \begin{cases} \sum_{i=1}^{N_2} \frac{24}{3V_2} \epsilon_{fw} \left(\left(\frac{\sigma_{fw}}{r_{wi}} \right)^{12} - 0.5 \left(\frac{\sigma_{fw}}{r_{wi}} \right)^6 \right) & r_{wi} < 2.5\sigma_{fw} \\ 0, & \text{otherwise} \end{cases} \tag{10}$$

By comparing Equations (8)–(10), it can observed that the multiplication factor 48 in Equation (8) is replaced by the factor 24 in Equations (9) and (10). This difference means that only 50 % of the fluid-wall interactions are taken into account at the calculation of the $\hat{p}_{\text{fluid}-\text{wall}}$ term [7]. In the MC simulations, Equations (7)–(10) yield instantaneous values from which ensemble averages are computed, i.e., $\langle \hat{p} \rangle$.

In all simulations, the LJ parameters of the fluid particles are $\epsilon_{\text{fluid}} = 1$, $\sigma_{\text{fluid}} = 1$, and the cut-off radius is $r_{\text{cut}} = 2.5\sigma_{\text{fluid}}$ for the fluid-fluid interactions. Regardless of the type of interaction potential used to calculate the interaction of the fluid particles and the wall in Simulation Box 2, the LJ parameter, $\sigma_{fw} = 1$ is used. In the case of the purely repulsive wall potential, the LJ parameter $\epsilon_{fw} = 1$ is used. In case of the attractive wall potential, five different values for the ϵ_{fw} LJ parameters are considered: $\epsilon_{fw} = 0.3, 0.5, 0.7, 1.0, 1.5$. In this study, all of the reported parameters are in dimensionless units. The LJ interactions are truncated and shifted (i.e., no tail corrections are applied). To avoid phase transitions, the temperature is fixed at $T = 2$ [31,32].

3. Theory

To investigate the difference between the differential and integral pressures, the ratio $\dfrac{d\langle\hat{p}\rangle}{dP}$ is calculated. The term $\dfrac{d\langle\hat{p}\rangle}{dP}$ is the ratio of the pressure gradient for mass transport in the bulk phase, $\dfrac{dP}{dL}$, and in the confined space $\dfrac{d\langle\hat{p}\rangle}{dL}$ (see Figure 3). Essentially, the ratio of driving forces, $\dfrac{d\langle\hat{p}\rangle}{dP}$, equals the ratio of transport coefficients when either P or $\hat{p}$ is used as driving force for mass transport in the corresponding transport equation. $\dfrac{d\langle\hat{p}\rangle}{dP}$ is referred to as the ratio of driving forces throughout this work. One possible approach to compute the ratio of driving forces is to perform simulations at different imposed differential pressures and calculate the difference in the differential and integral pressures. To avoid the necessity of performing several simulations, in this study the ratio of driving forces is calculated based on the fluctuation theory. Using this approach, the ratio of driving forces can be obtained by performing a single simulation.

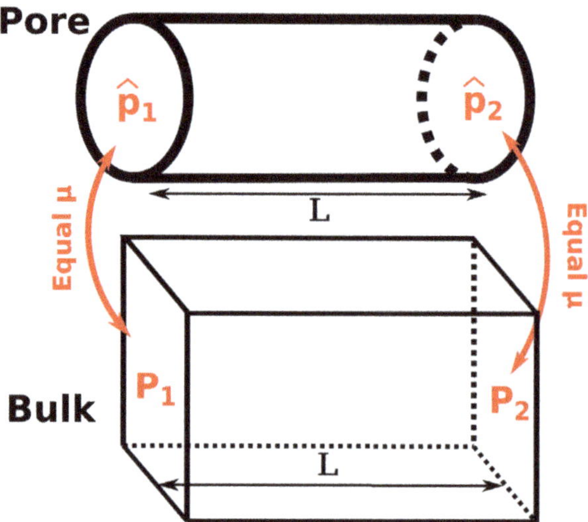

Figure 3. Schematic representation of a bulk fluid in equilibrium with a nanoconfined fluid in a pore. The concept of the ratio of driving forces for mass transport, $\dfrac{d\langle\hat{p}\rangle}{dP}$, can be introduced based on the definition of ratio of driving forces in the two systems, $\dfrac{d\langle\hat{p}\rangle}{dL}$ and $\dfrac{dP}{dL}$.

To obtain an expression for $\dfrac{d\langle\hat{p}\rangle}{dP}$, the partition function of the system is needed [23]:

$$Q = C \sum_{N_1=0}^{N_T} \frac{V_2^{N_T-N_1}}{N_1!\,(N_T-N_1)!} \int_0^{\infty} dV_1\, V_1^{N_1}\, e^{-\beta P V_1} \int dr^{N_T}\, e^{-\beta U}, \tag{11}$$

where C is a constant, $\beta = \frac{1}{k_B T}$, N_1 is the number of particles in Box 1, N_T is total number of particles in the simulation, V_1 is the volume of Box 1, V_2 is the volume of Box 2, and U is the potential energy. The ensemble average of a thermodynamic property X can be obtained using:

$$\langle X \rangle = \frac{\sum_{N_1=0}^{N_T} \frac{V_2^{(N_T-N_1)}}{N_1!\,(N_T-N_1)!} \int_0^{\infty} dV_1\, V_1^{N_1}\, e^{-\beta P V_1} \int dr^{N_T}\, e^{-\beta U} X}{\sum_{N_1=0}^{N_T} \frac{V_2^{(N_T-N_1)}}{N_1!\,(N_T-N_1)!} \int_0^{\infty} dV_1\, V_1^{N_1}\, e^{-\beta P V_1} \int dr^{N_T}\, e^{-\beta U}}. \tag{12}$$

Therefore, to obtain the expression for $\dfrac{d\langle \hat{p} \rangle}{dP}$, the following relation is used:

$$\frac{d\langle \hat{p} \rangle}{dP} = \frac{d}{dP} \frac{\sum_{N_1=0}^{N_T} \frac{V_2^{(N_T-N_1)}}{N_1!\,(N_T-N_1)!} \int_0^\infty dV_1\, V_1^{N_1}\, e^{-\beta P V_1} \int dr^{N_T} e^{-\beta U} \hat{p}}{\sum_{N_1=0}^{N_T} \frac{V_2^{(N_T-N_1)}}{N_1!\,(N_T-N_1)!} \int_0^\infty dV_1\, V_1^{N_1}\, e^{-\beta P V_1} \int dr^{N_T} e^{-\beta U}}. \tag{13}$$

By switching the order of the integration and differentiation, the following expression is obtained:

$$\frac{d\langle \hat{p} \rangle}{dP} =$$

$$\frac{\left(\sum_{N_1=0}^{N_T} \int_0^\infty dV_1 \int dr^{N_T} \frac{V_2^{(N_T-N_1)} V_1^{N_1}}{N_1!\,(N_T-N_1)!} (-\beta V_1) e^{-\beta P V_1} e^{-\beta U}\hat{p} \right)}{\left(\sum_{N_1=0}^{N_T} \int_0^\infty dV_1 \int dr^{N_T} \frac{V_2^{(N_T-N_1)} V_1^{N_1}}{N_1!\,(N_T-N_1)!} e^{-\beta P V_1} e^{-\beta U} \right)^2}$$

$$\times \left(\sum_{N_1=0}^{N_T} \int_0^\infty dV_1 \int dr^{N_T} \frac{V_2^{(N_T-N_1)} V_1^{N_1}}{N_1!\,(N_T-N_1)!} e^{-\beta P V_1} e^{-\beta U} \right)$$

$$- \frac{\left(\sum_{N_1=0}^{N_T} \int_0^\infty dV_1 \int dr \frac{V_2^{(N_T-N_1)} V_1^{N_1}}{N_1!\,(N_T-N_1)!} (-\beta V_1) e^{-\beta P V_1} e^{-\beta U} \right)}{\left(\sum_{N_1=0}^{N_T} \int_0^\infty dV_1 \int dr^{N_T} \frac{V_2^{(N_T-N_1)} V_1^{N_1}}{N_1!\,(N_T-N_1)!} e^{-\beta P V_1} e^{-\beta U} \right)^2}$$

$$\times \left(\sum_{N_1=0}^{N_T} \int_0^\infty dV_1 \int dr^{N_T} \frac{V_2^{(N_T-N_1)} V_1^{N_1}}{N_1!\,(N_T-N_1)!} e^{-\beta P V_1} e^{-\beta U} \hat{p} \right) \tag{14}$$

$$= \langle -\beta V_1 \hat{p} \rangle - ((\langle -\beta V_1 \rangle \langle \hat{p} \rangle)) = \beta(\langle V_1 \rangle \langle \hat{p} \rangle - \langle V_1 \hat{p} \rangle) = \frac{\langle V_1 \rangle \langle \hat{p} \rangle - \langle V_1 \hat{p} \rangle}{k_b T}.$$

The final expression in Equation (14) is essentially the cross correlation between V_1 and $\hat{p}$. In this study, this expression is used to calculate the ratio of driving forces.

4. Results and Discussion

In this work, the difference between the differential and integral pressure, $P - \langle \hat{p} \rangle$, and the ratio of driving forces $\dfrac{d\langle \hat{p} \rangle}{dP}$, are investigated. Two different pore geometries, a cylinder and a slit pore, are studied with varying fluid-wall interaction potentials. The effect of confinement is investigated for gas ($\rho \approx 0.1$) and liquid ($\rho = 0.58, 0.8$) phases, corresponding to $P = 0.2, 2.0, 6.0$, respectively.

4.1. Difference between the Differential and Integral Pressure

In Figure 4, the difference between the differential, P, and the ensemble average of the integral pressure, $\langle \hat{p} \rangle$, is shown as a function of the inverse radius, R^{-1}, of Box 2 for the cylindrical and slit pore cases for the two types of wall potentials. As can be seen in Figure 4, as R^{-1} decreases, the difference between the differential and integral pressure decreases in all cases. The data are fitted to $AR^{-1}+B$, where A and B are constants. The coefficient of determination of the fitted lines is above 0.99 showing that the relation between the R^{-1} and $P - \langle \hat{p} \rangle$ is indeed linear. For large radii ($R > 30\sigma$), where the fluid in Box 2 behaves like in the bulk, $P - \langle \hat{p} \rangle$ approaches 0, which is also indicated by the fitted lines. In Figure 4a,c,e, the difference between the two pressures are shown for the cylindrical pore. In Figure 4b,d,f, the difference in the pressures are shown for the slit pore. By comparing the magnitude of $P - \langle \hat{p} \rangle$ for the cylindrical and slit pore cases, it can be seen that the pressure difference is larger in the cylindrical pore than in the slit pore. The larger value of $P - \langle \hat{p} \rangle$ for the cylindrical pores can be attributed to the stronger confinement effects compared to the slit pore. The stronger confinement effects are also indicated by the steeper slopes (constant A in the fitted lines) of the cylindrical pore compared to the slit pore. It can also be observed that by increasing the interaction strength between the wall of Box 2 and the fluid, ϵ_{wf}, the difference between the differential and integral pressure at the same pore size decreases. By comparing the calculated values of $P - \langle \hat{p} \rangle$ for $P = 0.2$ and $P = 6.0$, it can be seen that the effect of the interaction strength between the wall in Box 2 and the fluid, ϵ_{wf}, is considerably larger at the lower differential pressure. For example, in the case of the slit pore, the ratio of slopes of the fitted line for $\epsilon_{wf} = 1.0$ and $\epsilon_{wf} = 1.5$ at $P = 0.2$ is 0.88, as shown in Figure 4b, and at $P = 6.0$ is 0.96, as shown in Figure 4f. This can be caused by the different number of particles inside the pore at the two differential pressures. In the case of $P = 0.2$, by increasing the interaction strength, more particles can enter the pore which results in larger integral pressures, while at $P = 6.0$, the pore is practically saturated for all ϵ_{wf}; therefore, the interaction strength has a lower effect on the integral pressure.

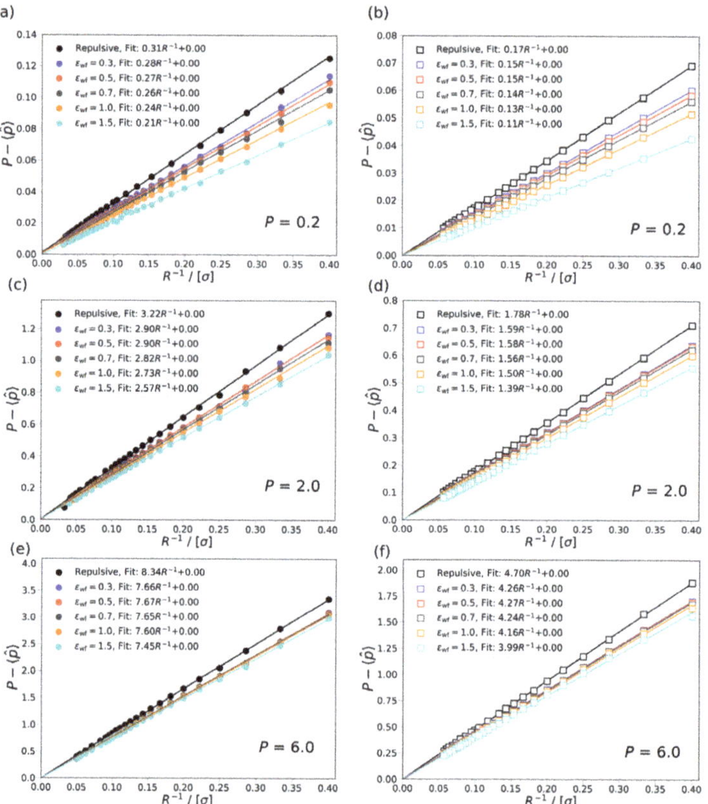

Figure 4. The difference between the differential, P, and the ensemble average of integral pressure, $\langle \hat{p} \rangle$, is shown as a function of the inverse radius R^{-1} of Box 2 at $P = 0.2$, 2.0, and 6.0 for cylindrical and slit pores with fluid-wall interactions. (**a,c,e**) The pressure difference is shown for cylindrical pores at differential pressure $P = 0.2$, 2.0, and 6.0, respectively. (**b,d,f**) The pressure difference is shown for slit pores at differential pressure $P = 0.2$, 2.0, and 6.0, respectively. The simulation results are shown with symbols, while the lines are fits to the data points. The equation used for the fitting is $AR^{-1} + B$, where A and B are constants. The colors denote the different level of attraction between the wall of Box 2 and the fluid, repulsive wall potential (black), $\epsilon_{wf} = 0.3$ (blue), 0.5 (red), 0.7 (gray), 1.0 (orange), and 1.5 (cyan). The results for the cylindrical pore are shown with closed circles and for the slit pores with open rectangles. The temperature of both boxes is set to $T = 2$. The average densities of Box 1 are $\rho \approx 0.10$, 0.58, and 0.8 at $P = 0.2$, 2.0, and 6.0, respectively.

Based on the work of Galteland et al. [11], the slope of the fitted line can also be related to the effective surface tension between the fluid particles and the wall in Box 2, i.e., $P - \langle \hat{p} \rangle \sim \frac{\hat{\gamma}^{fr}_{effective}}{R}$. The linear relation between $P - \langle \hat{p} \rangle$ and R^{-1} shows that the effective surface tension does not depend on the curvature of the wall. In Figure 4, it can be observed that the effective surface tension decreases as the fluid-wall interactions become more attractive. In Figure 4a–f, it can be seen that the ratio of the effective surface tensions (slopes of the fitted lines) for the same ϵ_{wf} with the slit and cylindrical pore is nearly constant. For example, at $\epsilon_{wf} = 0.5$ and $P = 2.0$, the ratio of the slope of the fitted lines for the slit and cylindrical pore is ~ 0.54, as shown in Figure 4c,d, and at $\epsilon_{wf} = 0.3$ and $P = 0.2$, the ratio of the two slopes is also ~ 0.54, as shown in Figure 4a,b. The ratio of the effective surface tensions between the slit and cylindrical pores are in the range of $\sim$0.52–0.56 and considered constant since it is within the uncertainties of simulations. The constant ratio of effective surface tension between the slit and cylindrical pore also indicates the larger confinement effects in the cylindrical pore.

4.2. Ratio of Driving Forces

In Figure 5, the ratio of driving forces for mass transport, $\frac{d\langle \hat{p} \rangle}{dP}$, is shown as a function of the inverse radius of Box 2, R^{-1}, at three different pressures, $P = 0.2, 2.0, 6.0$, for both the slit and cylindrical pores. From Figure 5, it can be observed that $\frac{d\langle \hat{p} \rangle}{dP}$ is considerably smaller than 1 for small pore sizes. This means that a change in the differential pressure, P, of Box 1 results in a smaller change in the integral pressure, $\langle \hat{p} \rangle$, of Box 2. This difference underlines the effect of the confinement on the pressure of the fluid in nanopores and shows the difference in driving forces in the bulk and confined fluid. As R increases, $\frac{d\langle \hat{p} \rangle}{dP}$ approaches 1, i.e., the fluid in the pore behaves more like a bulk fluid. As can be seen in Figure 5, $\frac{d\langle \hat{p} \rangle}{dP}$ is larger for the slit pore than for the cylindrical pore at the same conditions. This means that in case of the slit pore a change in the differential pressure, P, results in a larger difference in the integral pressure, $\langle \hat{p} \rangle$, than for the cylindrical pore. The larger change in the integral pressure indicates that the confinement effects are weaker in the slit pore. In Figure 5a,b, it can be observed that by increasing the interaction strength between the wall of the pore and the fluid particles, the ratio of the driving forces decreases. At higher differential pressures the decrease in the ratio of driving forces due to the increasing fluid-wall interaction strength becomes less pronounced, as shown in Figure 5c–f. The smaller influence of the interaction strength, ϵ_{fw}, on the ratio of driving forces may be caused by the increasing contribution of the fluid-fluid interactions to the integral pressure due to the larger number of fluid particles in the pore at higher differential pressure. In Figure 5a–f, the ratio of driving force is shown for the slit and cylindrical pores at the same differential pressure. It can be seen that the ratio of the slopes of the fitted lines for the same value of ϵ_{wf} with the slit and cylindrical pore is constant within the uncertainties of the simulations. The ratio of the slopes is in the range of $\sim$0.52–0.58, which indicates that the confinement effects in the cylindrical pore are almost twice as strong as in the slit pore.

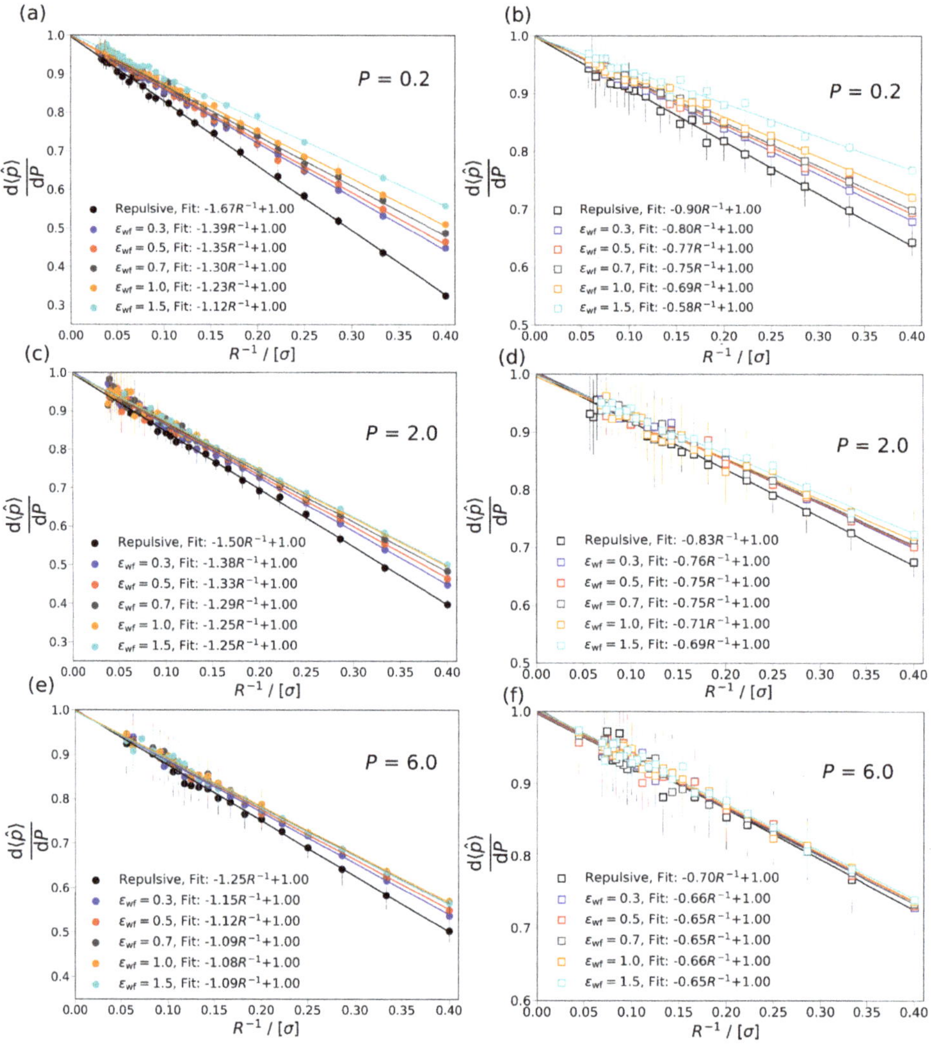

Figure 5. The ratio of driving forces, $\frac{d\langle\hat{p}\rangle}{dP}$, is shown as a function of the inverse radius, R^{-1}, of Box 2 at $P = 0.2, 2.0$, and 6.0 for the cylindrical and slit pore cases with repulsive and attractive wall potentials. (**a,c,e**) $\frac{d\langle\hat{p}\rangle}{dP}$ is shown for cylindrical pores at differential pressure $P = 0.2$, 2.0, and 6.0, respectively. (**b,d,f**) $\frac{d\langle\hat{p}\rangle}{dP}$ is shown for slit pores at differential pressures $P = 0.2$, 2.0, and 6.0, respectively. The simulation results are shown with symbols, while the lines are fits to the data points. The equation used for the fitting is $AR^{-1} + B$, where A and B are constants. The results for the cylindrical pore are shown with closed circles and for the slit pores with open rectangles. The colors denote the different wall potential used in Box 2, repulsive wall potential (black), $\epsilon_{wf} = 0.3$ (blue), $\epsilon_{wf} = 0.5$ (red), $\epsilon_{wf} = 0.7$ (gray), $\epsilon_{wf} = 1.0$ (orange), and $\epsilon_{wf} = 1.5$ (cyan). The temperature of both boxes is set to $T = 2$. The average densities of Box 1 are $\rho \approx 0.10, 0.58$, and 0.8 at $P = 0.2, 2.0$, and 6.0, respectively.

5. Conclusions

In this study, the new approach reported by Galteland et al. [11] is used to investigate the effects of confinement on a fluid in a nanopore by performing MC simulations. The simulations are carried out in a variation of the Gibbs ensemble with two simulation boxes in chemical equilibrium. One of the simulation boxes represents the bulk fluid with differential pressure P, and the other a slit or cylindrical pore with repulsive or attractive wall interaction potential. In case of the attractive wall potentials, several scenarios are considered for the strength of the interaction between the wall and the fluid particles. The effect of confinement is investigated for three differential pressures, $P = 0.2, 2.0, 6.0$, corresponding to gas ($\rho \approx 0.1$) and liquid phases ($\rho = 0.58, 0.8$). It is concluded that the difference between the differential and integral pressure $P - \langle \hat{p} \rangle$, for all studied cases, approaches 0 when $R \rightarrow \infty$. It is shown that the increase in the interaction strength between the wall and the fluid particles has smaller effect on the difference in the pressures, $P - \langle \hat{p} \rangle$, as the differential pressure increases. Based on the work of Galteland et al. [11], the difference of the differential and integral pressure is related to the effective surface tension between the fluid particles and wall of the pore. It is shown that the effective surface tension does not depend on the curvature of the wall. It is found that by considering a bulk fluid in equilibrium with a cylindrical or slit nanopore, the ratio of driving forces for mass transport in the bulk phase is larger than in the nanopore ($\frac{\mathrm{d}\langle \hat{p} \rangle}{\mathrm{d}P} << 1$) for small pore sizes. As R increases, $\frac{\mathrm{d}\langle \hat{p} \rangle}{\mathrm{d}P}$ approaches 1, i.e., the fluid in the pore behaves more like a bulk fluid. This clearly shows that the approximation that $\hat{p} \approx P$ does not hold on the nanoscale.

Author Contributions: T.J.H.V. and M.E. wrote the software, M.E. and O.G. carried out the simulations. S.K., D.B., and O.G. contributed to the conceptual design and the development of the work. The project was supervised by O.A.M., S.K., D.B., and T.J.H.V. All authors contributed to the writing of the manuscript and the scientific discussions. All authors have read and agree to the published version of the manuscript.

Funding: This research was funded by the Center of Excellence funding scheme of the Research Council of Norway, project no 262644, PoreLab. This work was also sponsored by NWO Exacte Wetenschappen (Physical Sciences) for the use of supercomputer facilities, with financial support from the Nederlandse Organisatie voor Wetenschappelijk Onderzoek (Netherlands Organisation for Scientific Research, NWO).

Acknowledgments: T.J.H.V. acknowledges NWO-CW (Chemical Sciences) for a VICI grant.

Conflicts of Interest: The authors declare no conflict of interest.

References

1. Gubbins, K.E.; Long, Y.; Śliwinska Bartkowiak, M. Thermodynamics of confined nano-phases. *J. Chem. Thermodyn.* **2014**, *74*, 169–183. [CrossRef]
2. Furukawa, H.; Cordova, K.; O'Keeffe, M.; Yaghi, O. The chemistry and applications of metal-organic frameworks. *Science* **2013**, *341*. [CrossRef] [PubMed]
3. Gu, Z.Y.; Yang, C.X.; Chang, N.; Yan, X.P. Metal-organic frameworks for analytical chemistry: From sample collection to chromatographic separation. *Accounts Chem. Res.* **2012**, *45*, 734–745. [CrossRef]
4. Yilmaz, B.; Müller, U. Catalytic applications of zeolites in chemical industry. *Top. Catal.* **2009**, *52*, 888–895. [CrossRef]
5. Glaser, R.; Weitkamp, J. The application of zeolites in catalysis. *Springer Ser. Chem. Phys.* **2004**, *75*, 159–212.
6. Erdős, M.; de Lange, M.F.; Kapteijn, F.; Moultos, O.A.; Vlugt, T.J.H. In Silico Screening of Metal–Organic Frameworks for Adsorption-Driven Heat Pumps and Chillers. *ACS Appl. Mater. Interfaces* **2018**, *10*, 27074–27087. [CrossRef]
7. Irving, J.H.; Kirkwood, J.G. The Statistical Mechanical Theory of Transport Processes. IV. The Equations of Hydrodynamics. *J. Chem. Phys.* **1950**, *18*, 817–829. [CrossRef]
8. Férey, G.; Serre, C. Large breathing effects in three-dimensional porous hybrid matter: Facts, analyses, rules and consequences. *Chem. Soc. Rev.* **2009**, *38*, 1380–1399. [CrossRef]
9. Song, H.; Yu, M.; Zhu, W.; Wu, P.; Lou, Y.; Wang, Y.; Killough, J. Numerical investigation of gas flow rate in shale gas reservoirs with nanoporous media. *Int. J. Heat Mass Transf.* **2015**, *80*, 626–635. [CrossRef]

10. Huber, P. Soft matter in hard confinement: phase transition thermodynamics, structure, texture, diffusion and flow in nanoporous media. *J. Phys. Condens. Matter* **2015**, *27*, 103102. [CrossRef]

11. Galteland, O.; Bedeaux, D.; Hafskjold, B.; Kjelstrup, S. Pressures Inside a Nano-Porous Medium. The Case of a Single Phase Fluid. *Front. Phys.* **2019**, *7*, 60. [CrossRef]

12. Todd, B.D.; Evans, D.J.; Daivis, P.J. Pressure tensor for inhomogeneous fluids. *Phys. Rev. E* **1995**, *52*, 1627–1638. [CrossRef]

13. Ikeshoji, T.; Hafskjold, B.; Furuholt, H. Molecular-level Calculation Scheme for Pressure in Inhomogeneous Systems of Flat and Spherical Layers. *Mol. Simul.* **2003**, *29*, 101–109. [CrossRef]

14. Walton, J.; Tildesley, D.; Rowlinson, J.; Henderson, J. The pressure tensor at the planar surface of a liquid. *Mol. Phys.* **1983**, *48*, 1357–1368. [CrossRef]

15. Blokhuis, E.M.; Bedeaux, D. Pressure tensor of a spherical interface. *J. Chem. Phys.* **1992**, *97*, 3576–3586. [CrossRef]

16. Wang, G.M.; Sevick, E.M.; Mittag, E.; Searles, D.J.; Evans, D.J. Experimental Demonstration of Violations of the Second Law of Thermodynamics for Small Systems and Short Time Scales. *Phys. Rev. Lett.* **2002**, *89*, 050601. [CrossRef]

17. Nilson, R.H.; Griffiths, S.K. Influence of atomistic physics on electro-osmotic flow: An analysis based on density functional theory. *J. Chem. Phys.* **2006**, *125*, 164510. [CrossRef]

18. Lee, J.W.; Nilson, R.H.; Templeton, J.A.; Griffiths, S.K.; Kung, A.; Wong, B.M. Comparison of Molecular Dynamics with Classical Density Functional and Poisson–Boltzmann Theories of the Electric Double Layer in Nanochannels. *J. Chem. Theory Comput.* **2012**, *8*, 2012–2022. [CrossRef]

19. Gjennestad, M.A.; Øivind Wilhelmsen. Thermodynamic stability of droplets, bubbles and thick films in open and closed pores. *Fluid Phase Equilibria* **2020**, *505*, 112351. [CrossRef]

20. Hill, T.L. *Thermodynamics of Small Systems*, 1st ed.; NY:Dover: New York, NY, USA, 1964.

21. Panagiotopoulos, A.Z. Direct determination of phase coexistence properties of fluids by Monte Carlo simulation in a new ensemble. *Mol. Phys.* **1987**, *61*, 813–826. [CrossRef]

22. Bai, P.; Siepmann, J.I. Selective adsorption from dilute solutions: Gibbs ensemble Monte Carlo simulations. *Fluid Phase Equilibria* **2013**, *351*, 1–6. [CrossRef]

23. Frenkel, D.; Smit, B. *Understanding Molecular Simulation*, 2nd ed.; Academic Press: London, UK, 2001.

24. Shi, W.; Maginn, E.J. Continuous Fractional Component Monte Carlo : An Adaptive Biasing Method for Open System Atomistic Simulations. *J. Chem. Theory Comput.* **2007**, *3*, 1451–1463. [CrossRef] [PubMed]

25. Poursaeidesfahani, A.; Torres-Knoop, A.; Dubbeldam, D.; Vlugt, T.J.H. Direct Free Energy Calculation in the Continuous Fractional Component Gibbs Ensemble. *J. Chem. Theory Comput.* **2016**, *12*, 1481–1490. [CrossRef] [PubMed]

26. Bennett, C.H. Efficient estimation of free energy differences from Monte Carlo data. *J. Comput. Phys.* **1976**, *22*, 245–268. [CrossRef]

27. Stecki, J. Steele (10-4-3) Potential due to a Solid Wall. *Langmuir* **1997**, *13*, 597–598. [CrossRef]

28. Jiménez-Serratos, G.; Cárdenas, H.; Müller, E.A. Extension of the effective solid-fluid Steele potential for Mie force fields. *Mol. Phys.* **2019**, *117*, 3840–3851. [CrossRef]

29. Weeks, J.D.; Chandler, D.; Andersen, H.C. Role of Repulsive Forces in Determining the Equilibrium Structure of Simple Liquids. *J. Chem. Phys.* **1971**, *54*, 5237–5247. [CrossRef]

30. Allen, M.P.; Tildesley, D.J. *Computer Simulation of Liquids*, 2nd ed.; Oxford University Press: Oxford, UK, 2017.

31. Smit, B. Phase diagrams of Lennard–Jones fluids. *J. Chem. Phys.* **1992**, *96*, 8639–8640. [CrossRef]

32. Potoff, J.J.; Panagiotopoulos, A.Z. Critical point and phase behavior of the pure fluid and a Lennard–Jones mixture. *J. Chem. Phys.* **1998**, *109*, 10914–10920. [CrossRef]

Article

Two-Phase Equilibrium Conditions in Nanopores

Michael T. Rauter [1,*], Olav Galteland [1], Máté Erdős [2], Othonas A. Moultos [2], Thijs J. H. Vlugt [2], Sondre K. Schnell [3], Dick Bedeaux [1] and Signe Kjelstrup [1]

[1]   PoreLab, Department of Chemistry, Norwegian University of Science and Technology, NO-7491 Trondheim, Norway; olav.galteland@ntnu.no (O.G.); dick.bedeaux@ntnu.no (D.B.); signe.kjelstrup@ntnu.no (S.K.)

[2]   Engineering Thermodynamics, Process and Energy Department, Delft University of Technology, Leeghwaterstraat 39, 2628CB Delft, The Netherlands; m.erdos-2@tudelft.nl (M.E.); o.moultos@tudelft.nl (O.A.M.); T.J.H.Vlugt@tudelft.nl (T.J.H.V.)

[3]   Department of Materials Science and Engineering, Norwegian University of Science and Technology, NO-7491 Trondheim, Norway; sondre.k.schnell@ntnu.no

*   Correspondence: michael.t.rauter@ntnu.no

Received: 25 February 2020; Accepted: 21 March 2020; Published: 26 March 2020

Abstract: It is known that thermodynamic properties of a system change upon confinement. To know how, is important for modelling of porous media. We propose to use Hill's systematic thermodynamic analysis of confined systems to describe two-phase equilibrium in a nanopore. The integral pressure, as defined by the compression energy of a small volume, is then central. We show that the integral pressure is constant along a slit pore with a liquid and vapor in equilibrium, when Young and Young–Laplace's laws apply. The integral pressure of a bulk fluid in a slit pore at mechanical equilibrium can be understood as the average tangential pressure inside the pore. The pressure at mechanical equilibrium, now named differential pressure, is the average of the trace of the mechanical pressure tensor divided by three as before. Using molecular dynamics simulations, we computed the integral and differential pressures, $\hat{p}$ and p, respectively, analysing the data with a growing-core methodology. The value of the bulk pressure was confirmed by Gibbs ensemble Monte Carlo simulations. The pressure difference times the volume, V, is the subdivision potential of Hill, $(p - \hat{p})V = \epsilon$. The combined simulation results confirm that the integral pressure is constant along the pore, and that ϵ/V scales with the inverse pore width. This scaling law will be useful for prediction of thermodynamic properties of confined systems in more complicated geometries.

Keywords: pressure; confinement; equilibrium; thermodynamic; small-system; hills-thermodynamics; pore; nanopore; interface

1. Introduction

It is well known that thermodynamic properties of fluids significantly change when the fluid phases become confined [1,2]. Given the importance of porous media for applications in for example fuel cells [3], batteries [4], and membranes for drinking water production [5–7], it is essential to understand how properties change upon confinement. In particular, one would like to be able to predict how thermodynamic properties of the system change with size. We have earlier documented how the thermodynamic factors and the pressures of some systems scale with a characteristic inverse system size (surface area over volume) [8–10]. The aim of the present work is to continue this line of work and present a geometric scaling law for the pressure of a nanopore. This work can, in a general context, be seen as an extension of Hill's thermodynamics of small systems [11]. The essential idea of Hill was to introduce a large ensemble of $\mathcal{N}$ small systems, enabling the use of standard thermodynamic tools for the ensemble. System properties were obtained by dividing the ensemble value by $\mathcal{N}$. In this way, he was able to deal with the impact of shape- and size-variation of the small system, within the normal structure of thermodynamics. The dimensions of a nano-pore, which make molecular interactions at the pore surface important, can promote changes in phase transitions, surface adsorption, pore vapor pressure, and phase stability [2,12–14]. The literature offers, however, several definitions of the pressure inside a nano-porous material [15,16]. This situation calls for a robust definition of the representative elementary volume (REV) of the small system, to serve as a basis for a definition of pressure. Kjelstrup et al. [17] offered a definition of the REV-pressure based on the compression energy of the REV. Galteland et al. [18] showed, by evaluating this property for a single component in a nanoporous system, that the REV was not sufficiently described by the bulk fluid pressure. Also the integral pressure, $\hat{p}$, as defined by Hill was required to describe the degree of confinement in the REV. In Hill's [11] terminology, one distinguishes between the integral pressure $\hat{p}$, and the differential pressure, p. Erdos et al. [10] provided an expression for the ratio of the driving forces for adsorption into a pore, by means of the integral and differential pressures. They expressed the ratio by the cross-correlation of the volume and the integral pressure.

Hill defined the difference of the two pressures times the volume by the subdivision potential $\epsilon = (p - \hat{p})V$ [11]. The integral pressure can differ significantly from the differential pressure in confined systems [10,18]. In the limit of a bulk fluid, they are the same, however. Galteland et al. studied pores between an fcc type-lattice of spheres [18], and found that the difference in the differential and integral pressure depended on the inverse radius of the spheres in the lattice. Here, we investigate a slit pore of nano-scale dimensions, with a constant pore slit-width ranging from 3 to 5 nm, and filled with a vapor and liquid phase of the same component. The purpose is to find thermodynamic and mechanical equilibrium conditions in terms of the pressures defined by Hill, and compare these to existing descriptions in terms of Young's and Young–Laplace's laws. The analysis leads to a new equilibrium condition, stated in terms of the integral pressure: This property is constant along the nano-sized pore. When this condition is combined with Youngs' equation for the contact angles, one recovers the Young–Laplace equation.

For the geometry studied here, it is possible to derive a scaling law for the integral pressure. The pore shape is kept constant in the derivations, as shape is a variable in small system thermodynamics [19]. The new scaling law relates properties of one system size to another size, with the same shape. This is helpful in a situation where measurements are lacking. Hill extended the thermodynamic variable set, including an ensemble of the small system, thereby providing a new basis for introduction and examination of size effects in thermodynamics. Hill's formulation of surface thermodynamics contain Gibbs' method [20].

Molecular dynamics and Monte Carlo simulations [10,21] are excellent tools for equilibrium studies of confined systems. Two fluid phases will be simulated in a slit pore of varying width. Focus is directed to the equilibrium conditions, when single phases are in contact with each other and the wall. The choice of the system was motivated by the simple geometry, which makes it possible to test particular scaling laws for slit pores and cylindrical pores.

The structure of the paper is as follows: We first present the equations of a REV leading to the relation between the differential pressure p and the integral pressure $\hat{p}$. We next test the derived equations for slit pores of different sizes and different interactions between pore wall and fluid. The new growing-core methodology, brings out the difference between the differential and the integral pressure. Finally, results are discussed and put in perspective.

2. Theory

2.1. Thermodynamic Relations for Small Systems

Hill used the name "small system" for systems with properties that are not extensive in the system volume [11]. Such systems are frequently confined, i.e., bounded by the surroundings. To obtain a generalized thermodynamic description of small systems, and to deal with size and shape-dependencies, Hill [11] introduced an ensemble of $\mathcal{N}$ identical and independent replicas of the small system. Keeping the total entropy S_t, and particle number, N_t, constant, while increasing the number of small systems $\mathcal{N}$ (each with constant volume V) in the ensemble, an extended version of the Gibbs equation was obtained:

$$dU_t = TdS_t - p\mathcal{N}dV + \sum_j \mu_j dN_{j,t} - \widehat{p}Vd\mathcal{N} \tag{1}$$

which we call the Hill-Gibbs equation. Here, U_t is the total internal energy of the ensemble, T is the temperature, p is the pressure, and μ_j is the chemical potential of component j. The new term, $-\widehat{p}Vd\mathcal{N}$, is the reversible work needed to change the total volume, by changing the number of small systems, keeping S_t, V, and $N_{t,j}$ constant.

Hill [11] distinguished between p and $\widehat{p}$ by calling the properties the differential and integral pressure, respectively. This work aims to elucidate the difference of these variables for a small system which is confined to a pore, in the grand-canonical ensemble.

The average values in the grand-canonical ensemble for a single small system are

$$U = \frac{U_t}{\mathcal{N}}, \quad N_j = \frac{N_{j,t}}{\mathcal{N}}, \quad S = \frac{S_t}{\mathcal{N}}. \tag{2}$$

By introducing these equations into Equation (1), we obtain the usual Gibbs equation for dU, and what we call the Hill-Gibbs-Duhem's equation.

$$d(\widehat{p}V) = SdT + \sum_j N_j d\mu_j + pdV \tag{3}$$

The equation reduces to the usual Gibbs-Duhem equation when $\widehat{p} = p$. When T and μ_j are constant in Equation (3), we obtain a relation between the integral and differential pressure when the shape of the system is controlled [11]:

$$p = \left(\frac{\partial \widehat{p}V}{\partial V} \right)_{T,\mu} = \widehat{p} + V \left(\frac{\partial \widehat{p}}{\partial V} \right)_{T,\mu}. \tag{4}$$

The expression shows that p and $\widehat{p}$ differ for small systems, for which $\widehat{p}$ depends on the volume. In a macroscopic system, with $V \to \infty$, p and $\widehat{p}$ are the same. The relation clarifies the chosen names; p is obtained by differentiation.

The above equations will be applied to a two-phase system in a slit pore and a cylindrical pore. Inside the pore, the differential pressure of the liquid differ from that of the vapor phase. For the time being, we introduce the notation p_d^v and p_d^l for the differential pressure of the vapor and liquid, respectively, and p_d^f to indicate either of the two. The difference of the differential and the integral pressure times the volume is equal to the subdivision potential [11], ϵ. By rearranging Equation (4), using $p_d^f = p$, we obtain:

$$p - \widehat{p} = \frac{\epsilon(T, V, \mu_j)}{V} = V \left(\frac{\partial \widehat{p}}{\partial V} \right)_{T,\mu_j}. \tag{5}$$

The subdivision potential is an additive thermodynamic property [11]. When such a property is divided by the small system volume, a scaling law can be found; of a similar type as we have found earlier [8,9]. The ratio ϵ/V is proportional to the the inverse characteristic size Ω/V of the system, where Ω is the surface area between the fluid and the solid [18]. In general terms,

$$p - \widehat{p} = \frac{\epsilon}{V} = \zeta^\infty + \zeta^s \left(\frac{\Omega}{V} \right) + \zeta^{se} \left(\frac{\Omega}{V} \right)^2 + ... \tag{6}$$

The coefficients ζ^∞, ζ^s and ζ^{se} do not depend on Ω/V, but are functions of T and μ_j. As p and $\widehat{p}$ are equal in the thermodynamic limit, it follows that the coefficient $\zeta^\infty = 0$.

2.2. The Integral Pressure of a Representative Volume Element

To define the differential and integral pressures, and the conditions for thermodynamic and mechanical equilibrium between two phases within a pore, we need suitable REVs. In a porous medium with two fluid phases, we need more than one type of REV, one for each phase. A REV needs, as the name says, to be representative for all molecular interactions in the system [17]. It should be large enough to contain a statistically representative amount of the fluids and solids. The purpose of the REV is to define a volume element, to be used to define equilibrium, but also to define local equilibrium in a large system where fluid transports take place [17].

A liquid droplet in equilibrium with its vapor inside a slit pore is an example of a two-phase confined fluid. The droplet system is illustrated in Figure 1. Various REVs are possible. Each REV need to cover the whole cross section of the pore. In the following we are concerned with two REVs, one in the vapor- and one in the liquid phase. None of them includes the surface between liquid and vapor. Two examples of REVs (REV$_1$ and REV$_2$) are illustrated in Figure 1. The x-axis is located at the center of the pore.

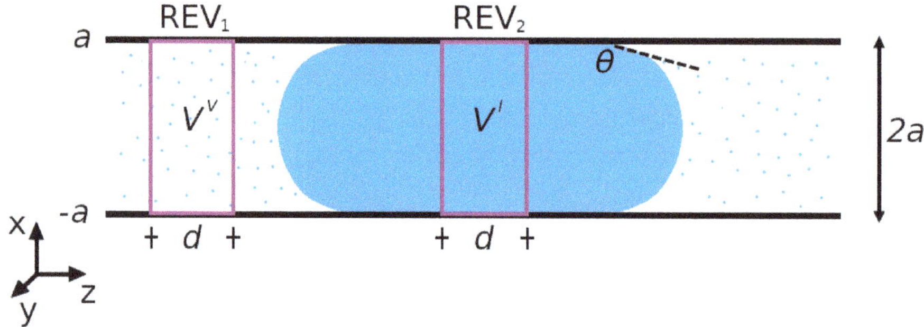

Figure 1. Slit pore of width 2*a* with a liquid droplet in the middle, which is in equilibrium with a vapor phase on both sides. The representative volume elements are REV$_1$ (vapor of volume V^v) and REV$_2$ (liquid of volume V^l), both with length *d*. The *x*-axis is located at the center of the pore.

The integral pressure of an isothermal REV with constant chemical potentials is expressed in terms of the compression energy, $\hat{p}V$, which is defined by the grand potential, Y [11].

$$\hat{p}V = -Y \equiv k_{\mathrm{B}}T \ln \Xi \tag{7}$$

Here k_{B} is the Boltzmann constant and Ξ the grand-canonical partition function. Similar to ϵ, also the compression energy is additive. It is the sum of contributions from all phases and interfaces present within the REV. The grand potential provides then the statistical mechanical link to particle behaviour, and is a tool for pressure computations (see below). The system is small in the sense of Hill, when the integral pressure deviate from the (bulk) fluid pressure.

The general expression for the compression energy for a REV$_n$, with $n = 1, 2$ is then

$$\hat{p}_{\mathrm{REV}_n} V = \hat{p}^f V^f - \hat{\gamma}^{fs} \Omega^{fs} \tag{8}$$

where V is the volume of the REV, V^f is the volume of the fluid, Ω^{fs} is the surface area between the fluid and the wall, and $\hat{\gamma}^{fs}$ is the integral surface tension at the wall, or the surface tension of the small system. In the present case, the two REVs are completely filled with a fluid; a vapor (REV$_1$) or a liquid (REV$_2$). The REV volume in both cases obeys $V = V^f = V^v = V^l = 2adw$, where a is the slit pore half width, d the REV length and w the width of the REV. Superscripts v and l refer again to the vapor and the liquid. The surface area between fluid and solid is $\Omega^{fs} = 2dw$. The surface tension between fluid and pore surface are $\widehat{\gamma}^{ls}$ and $\widehat{\gamma}^{vs}$, respectively, where the hat indicate that the system may be small. In that case, the surface tension is a function of the contact area/system geometry.

For sufficiently large values of a, the fluid in the center of the pore has macroscopic bulk properties. In that case; the integral pressure in the vapor $\widehat{p}^v$ and the liquid $\widehat{p}^l$ are independent of V^v or V^l, respectively, and

$$\widehat{p}^v = p^v, \qquad \widehat{p}^l = p^l. \tag{9}$$

Here, p^l and p^v are bulk pressures of the liquid and vapor, respectively.

In a slit pore, the wall surface is not curved. When the distance between the walls ($2a$) is sufficiently large, the fluid–solid surface tension show macroscopic behaviour, and:

$$\widehat{\gamma}^{vs} = \gamma^{vs}, \qquad \widehat{\gamma}^{ls} = \gamma^{ls}. \tag{10}$$

By introducing the expressions, Equations (9) and (10), the volume and the fluid–solid surface area into Equation (8), we obtain expressions for the integral pressure of the liquid and vapor REVs:

$$\widehat{p}_{\text{REV}_2} = p^l - \frac{\gamma^{ls}}{a}, \qquad \widehat{p}_{\text{REV}_1} = p^v - \frac{\gamma^{vs}}{a}. \tag{11}$$

Suppose now that integral pressure is constant everywhere at equilibrium, or that:

$$\widehat{p}_{\text{REV}_2} = \widehat{p}_{\text{REV}_1}. \tag{12}$$

This means we can write:

$$p^l = p^v + \frac{\gamma^{ls} - \gamma^{vs}}{a} \tag{13}$$

Young's law derives from a force balance at phase equilibrium:

$$\gamma^{ls} = \gamma^{vs} + \gamma^{lv} \cos \theta, \tag{14}$$

Here γ^{lv} is the liquid-vapor surface tension. We introduce this law in the equation above and obtain:

$$p^l = p^v + \frac{\gamma^{lv}}{a} \cos \theta = p^v + \frac{\gamma^{lv}}{R}. \tag{15}$$

This is Young–Laplace's equation for a slit pore. Young–Laplace's equation has been verified in the experimental literature numerous times [22,23]. The radius of curvature of the liquid-vapor surface, $R = a/\cos \theta$, was used, where θ is the contact angle at the pore wall, see Figure 1. We have found this equation without examining the liquid-vapor interface per se.

Equation (12) is thus equivalent to validity of the laws of Young and Young–Laplace. These familiar laws apply to thermodynamic and mechanical equilibrium, so we expect the equality to do the same. In this sense we have derived a new way to express phase equilibrium in nanopores.

Consider again a state of equilibrium where the integral pressure is constant. The integral pressures of the fluids are shown by Equation (11). For each fluid, the difference of the differential and integral pressure for the slit pore follows from Equation (4):

$$p - \widehat{p} = a^3 \left(\frac{\partial \widehat{p}}{\partial a^3} \right)_{T,\mu} = \frac{\gamma^{fs}(\alpha)}{3a}. \tag{16}$$

where we have indicated that the surface tension depends on a parameter α which characterizes the molecular interactions of the fluid and the wall. The shape is a variable in small system thermodynamics, and remains constant when we determine the derivative. To keep the shape constant, means to take the isomorphic derivative of the integral pressure in Equation (16).

A change from the slit pore to a cylindrical pore, will change the proportionality factor in a concrete way. The ratio of the area over the volume in the REV of a cylindrical pore is double the value in a slit pore. The ratio of the slope in a cylindrical pore over that from a slit pore is therefore equal to 2. The corresponding scaling law for a cylindrical pore becomes:

$$p - \hat{p} = a^3 \left(\frac{\partial \hat{p}}{\partial a^3} \right)_{T,\mu} = \frac{2\gamma^{fs}(\alpha)}{3a}. \tag{17}$$

By changing the geometry of the pore only, keeping the fluid wall interactions constant, we predict that the slope of pressure difference vs the surface tension will change by a factor 2.

For large systems (when $a \rightarrow \infty$), the difference between p and $\hat{p}$ disappears, leading to $p = \hat{p} = p^f$ in the REV. For small systems, however, the difference between p and $\hat{p}$ is described by the ratio between fluid–solid surface tension and the half channel width (γ^{fs}/a) divided by 3. We have thus found that the pressure difference follows a scaling law, similar to the one proposed by the Small System Method [8,9], and given in Equation (6).

Equation (16) can be tested by molecular dynamics simulations. In such a test, one can identify independently all terms in the equation, and confirm that the condition $\hat{p}_{REV_1} = \hat{p}_{REV_2}$ holds. Essential in the calculation is that p and $\hat{p}$ refer to the same volume.

These theoretical results demonstrate why it may pay to take the effort to invoke the perhaps more complicated thermodynamic scheme of Hill. Young and Young–Laplaces' laws describe the same physical phenomenon as Equations (16) and (17). Hill's theory shows how the two Equations (16) and (17) are linked to each other and to new variables (e.g., ϵ), in manners which can be tested.

2.3. A Mechanical Interpretation of the Pressures

The above derivation showed that the condition of constant integral pressure at equilibrium is required by Young's and Young–Laplace's laws. The integral and differential pressures are therefore also related to the mechanical pressure tensor. The mechanical pressure tensor will from now be denoted with upper case P, while the thermodynamic pressures are denoted as before, with lower case p. A complete set of relations between P and p is provided in the Appendix A. Here, we present the main relations.

Consider again the liquid droplet in a slit pore, as shown in Figure 1. The slit pore wall is parallel to the yz-plane. The pressure tensor component acting normal to the wall is the xx-component, $P_N = P_{xx}$ (cf. the Appendix A). There are two tangential components of the pressure tensor, the yy-component and the zz-component, which are equal in equilibrium. The tangential pressure tensor component is the average of the two, $P_T = \frac{1}{2}(P_{yy} + P_{zz})$. The surface tension is minus the excess of the tangential pressure tensor [24],

$$\gamma^{fs} = \frac{1}{2} \int_{-a}^{a} (P_N(x) - P_T(x)) \mathrm{d}x. \tag{18}$$

The factor of one half accounts for the fact that the integral is for two fluid–solid surfaces. By introducing the fluid–solid surface tension into the equation for the integral pressure of a single fluid phase, cf. Equation (11), we obtain

$$\hat{p} = p^f - \frac{1}{2a} \int_{-a}^{a} (P_N(x) - P_T(x)) \mathrm{d}x, \tag{19}$$

where superscript f again refers to either the liquid or vapor. The normal component of the mechanical pressure tensor is constant through flat surfaces at mechanical equilibrium. The integral of the normal component divided by $2a$ is therefore equal to P_N. We assume that the normal component is equal to the differential pressure of the bulk fluid, $p^f = P_N$. Below, the assumption is shown to hold.

The differential pressure of the bulk fluid and the integral over the normal component of the pressure cancel each other, and the integral pressure is equal to the average tangential component,

$$\widehat{p} = \frac{1}{2a} \int_{-a}^{a} P_T(x)\mathrm{d}x = \langle P_T \rangle. \tag{20}$$

This equation shows that the integral pressure of a bulk fluid in a slit pore at mechanical equilibrium can be understood as the average tangential pressure.

The differential pressure can also be related to the mechanical pressure tensor. The difference between the differential and integral pressures is equal to the fluid–solid surface tension divided by $3a$. By introducing Equation (18) for the fluid–solid surface tension, we obtain

$$p - \widehat{p} = \frac{1}{6a} \int_{-a}^{a} (P_N(x) - P_T(x))\mathrm{d}x \tag{21}$$

By reorganizing and inserting the average of the tangential pressure for the integral pressure,

$$p = \frac{1}{3} \left(\langle P_N \rangle + 2\langle P_T \rangle \right) = \frac{1}{3}\mathrm{Tr}\langle P \rangle \tag{22}$$

The differential pressure is equal to the average of the trace of the mechanical pressure tensor divided by 3.

3. Simulations

3.1. Molecular Dynamic Simulations

3.1.1. System

Simulations were carried out for a droplet in a slit pore. The droplet in the middle of the pore was surrounded by its vapor phase on both sides (see Figure 2). The simulation box, with the side lengths $L_x = L_y \neq L_z$ had periodic boundary conditions in all directions.

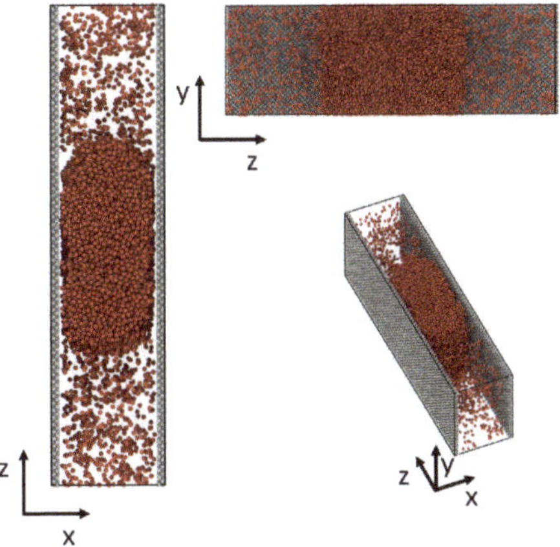

Figure 2. System used in the simulation. The droplet in the middle of the slit pore is surrounded by a vapor phase on both sides.

The simulation box was divided into n rectangular layers along the z-axis with a layer thickness of $\Delta z = L_z/n$. Both the pore-wall (grey) and the fluid (red) consisted of Lennard-Jones/spline particles [25]. The particles of the wall were immobilized, such that the particles were able to interact with the fluid particles, but not move. Due to periodic boundary conditions, the wall thickness was

$t = 3.5\sigma_0$ and had a density of $\rho_{wall} = 1.05$. The fluid inside the pore was initialized in a way that the overall density inside the pore was $\rho_{fluid} = 0.27$, thus in the two phase regime [25]. The averaged density of the bulk liquid was $\rho_{liquid} = 0.75 \pm 0.0015$ and the one of the vapor $\rho_{vapor} = 0.03 \pm 0.0025$.

The simulations were carried out using LAMMPS [26] in the canonical (NVT) ensemble at a temperature of $T^* = 0.70$. The Nosé-Hoover thermostat [27] was used to keep the temperature constant. We used Lennard-Jones reduced units, which are indicated by superscript * [28]. The system was simulated for four slit widths, $a = 8.7\sigma_0$, $10.45\sigma_0$, $12.2\sigma_0$ and $13.7\sigma_0$. The half channel widths were chosen so that both, the pressure of the fluid in the center, p^f, as well as the fluid–solid surface tension, γ^{fs}, are independent of the distance a, i.e., they show macroscopic bulk properties. Furthermore, the slit widths are large enough to avoid the effect of the disjoining pressure, as discussed in [13].

3.1.2. Particle Interaction Potential

The interaction between particles of type i and j was defined by the Lennard-Jones/spline potential [25,28],

$$u_{ij}(r) = \begin{cases} 4\varepsilon_{ij}\left[\left(\frac{\sigma_{ij}}{r}\right)^{12} - \alpha_{ij}\left(\frac{\sigma_{ij}}{r}\right)^{6}\right] & \text{if } r < r_{s,ij} \\ a_{ij}(r - r_{c,ij})^2 + b_{ij}(r - r_{c,ij})^3 & \text{if } r_{s,ij} < r < r_{c,ij} \\ 0 & \text{if } r > r_{c,ij} \end{cases} \tag{23}$$

where r is the distance between the particles. The interaction parameters, ε_{ij} and σ_{ij}, were set to 1 for all particle pairs (Lennard-Jones units). The interaction parameter α_{ij} is used to control the attractive interaction between the wall and the fluid, where we used $\alpha_{sf} = 0.05$, 0.15 and 0.25. For wall-wall and fluid-fluid interactions the interaction parameter was set to $\alpha_{ss} = \alpha_{ff} = 1$. All α_{sf} values represent a strong, repulsive interaction between wall and fluid, thus a non-wetting behaviour. The parameters a_{ij}, b_{ij}, $r_{c,ij}$ and $r_{s,ij}$ were determined such that the potential and its derivative are continuous functions at $r_{s,ij} = 1.24\sigma$ and $r_{c,ij} = 1.74\sigma$ [25].

3.1.3. Computation of REV Pressures and Wall-Fluid Surface Tension

The mechanical pressures in any layer l or combination of layers, or in a REV (core plus all layers included), were computed following Kirkwood [24]:

$$P_{\beta\kappa} = \frac{1}{3V}\sum_{i\in l}m_i(v_{i,\beta} - v_{m,\beta})(v_{i,\kappa} - v_{m,\kappa}) - \frac{1}{6V}\sum_{i\in l}\sum_{j=1}^{N}(r_{ij,\beta}f_{ij,\kappa}). \tag{24}$$

The subscripts β and κ denote the Cartesian coordinates, x, y and z, while V is the volume of the layer, m_i is the mass of particle i, $v_{i,\beta}$ is the velocity of fluid particle i in the β-direction, and $v_{m,\beta}$ is the average velocity in the β-direction. This velocity is zero in our equilibrium simulations. The properties $r_{ij,\beta}$ and $f_{ij,\beta}$ are, respectively, the distance and the force between particle i and j in the β-direction. The first term is the kinetic energy contribution from the particles, and the second term arise from pairwise interactions. Because of isotropy, the tensor is diagonal (see Appendix A).

Half the value of the pairwise energy contribution is assigned to the layer that contains particle i, while the other half is assigned to the layer that contains particle j. A more precise distribution of the contributions is available [29–32], but the present method has sufficiently small errors for the volumes used.

The pressure of a fluid volume element was calculated, starting with the core of the pore (lightest color in Figure 3), which has volume V_1. Volume was gradually added to the core, including in the end also the wall (black color in Figure 3). In this manner, the volume of the core was growing, layer by layer, until in the end $s - 1$ layers were added to the central core. The pressure was computed for the core (V_1), for the core plus one subsequent layer (V_2), and for the total volume (V_{tot}). The total volume, V_{tot}, is equal to the volume V of the REV (see Figure 1). Thus, the pressure was computed in growing pore core -volumes until the volume was large enough to cover the whole width (2*a*) of the pore. This new growing-core methodology enabled an calculation of the pressure in the center of the pore (the core) and the averaged pressure of the whole cross section of the pore, including all interactions with the pore wall. When all layers plus the wall inside the pore were included (V_{tot}), we obtained $\hat{p}$ of the REV. As long as there is no impact of the wall, (the core plus layers is sufficiently far away from the wall) we observed the bulk fluid pressure, p^f. This procedure was applied to two REVs, one in the liquid and one in the vapor phase.

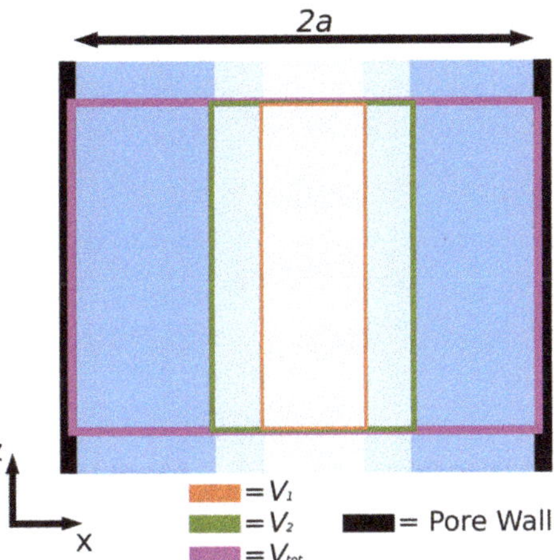

Figure 3. The core layer of the slit pore (lightest color) and subsequent layers (progressively darker color) used in the calculation of the pressure tensor of the fluid. V_1 and V_2 are the volumes of the core and the core plus one layer, respectively. V_{tot} is the total volume, covering all interactions with the pore wall.

The fluid–solid surface tension was computed as the excess of the tangential pressure tensor component [24] from Equation (18).

The extrapolated normal component P_N was taken to be equal to the pressure in core, where the fluid–solid surface has no effect on the pressure. The tangential component of the pressure was calculated using $P_T = \frac{1}{2}(P_{yy} + P_{zz})$. The value was obtained using a low resolution in x-direction and by averaging over multiple chunks of the liquid phase, along the pore in the z-direction.

3.2. Gibbs Ensemble Monte Carlo

In the derivation of Equation (20), the pressure tensor normal to the slit pore wall, P_N, is assumed to be equal to the bulk pressure p^f. This assumption was tested using Gibbs ensemble Monte Carlo simulations for two values of p^f (0.2 and 6.0 reduced units). The method is the same as reported recently by Erdős et al. [10]. In the simulations, two simulation boxes were defined. Simulation box 1 represents a reservoir of a bulk fluid with periodic boundary conditions applied in all directions. The fluid particles interact with the shifted and truncated Lennard-Jones potential with $\sigma = 1, \epsilon = 1$ and cut-off $r_c = 2.5$. The pressure, p^f, and temperature, T, of the fluid in simulation box 1 are imposed, while the number of particles, N_1, and the volume of the box can fluctuate. Simulation box 2 was a slit pore with fixed volume and temperature. The size of the slit pore was defined by the distance between the two parallel planes, 2*a*. In simulation box 2, periodic boundary conditions were applied in the directions parallel to the planes. The total number of particles, $N_1 + N_2$, was fixed in the simulations. Simulation box 1 and 2 can exchange particles, which ensured that the two boxes were in chemical equilibrium. The pressure of the fluid in simulation box 2 was equal to the differential pressure in equilibrium with the pressure imposed in the reservoir (box 1), p^f. A more detailed description of the simulation setup is given by Erdős et al. [10]. The high and low pressures were computed for varying pore sizes a.

4. Results and Discussion

The results for the independent computations of the pressure of a bulk fluid and corresponding pressure in the pore, using Gibbs ensemble Monte Carlo simulations, are shown in Table 1 in Section 4.1. The results of the molecular dynamics simulations are shown in Figures 4–7 in Sections 4.2–4.5. The figures show the pressure variation from the center of the slit to the wall, in the direction perpendicular to the wall (Figure 4) and parallel to the wall (Figure 5). The next figures confirm, as we shall see, that we can use a constant $\hat{p}$ to define equilibrium, and that a small system scaling law applies.

4.1. Pressure Component Normal to the Pore Wall

The normal component of the pressure tensor in simulation box 2 and bulk pressure in simulation box 1 are shown in Table 1.

Table 1. The pressure tensor normal to the slit pore wall, P_N, and the bulk pressure p^f with varying pore sizes a. Two total pressures are studied for various pore sizes.

Pore Size a/σ_0	Normal Pressure, $P_N \sigma_0^3/\varepsilon_0$	Bulk Fluid Pressure, $p^f \sigma_0^3/\varepsilon_0$
5	6.178 ± 0.005	6.023 ± 0.005
6	6.017 ± 0.005	6.011 ± 0.005
9	6.014 ± 0.005	5.995 ± 0.006
12	6.011 ± 0.005	5.995 ± 0.003
15	6.029 ± 0.004	5.995 ± 0.004
21	6.011 ± 0.004	6.000 ± 0.003
27	5.997 ± 0.004	6.002 ± 0.003
35	5.998 ± 0.004	5.997 ± 0.003
5	0.2011 ± 0.0009	0.2002 ± 0.0002
6	0.1988 ± 0.0009	0.2005 ± 0.0002
9	0.2007 ± 0.0008	0.2000 ± 0.0002
12	0.2020 ± 0.0007	0.2002 ± 0.0002
15	0.1980 ± 0.0007	0.1997 ± 0.0002
21	0.1993 ± 0.0006	0.1997 ± 0.0001
27	0.2005 ± 0.0006	0.2000 ± 0.0001
35	0.2000 ± 0.0005	0.2001 ± 0.0001

The pressure tensor normal to the pore wall in simulation box 2 is according to Table 1 equal to the bulk pressure in simulation box 1 within the accuracy of the simulation. The results show that this assumption is correct. Therefore, the integral pressure can be understood as the average value of the pressure tensor components tangential to the pore wall, $\hat{p} = \langle P_T \rangle$.

4.2. Pressure Variation in the Direction Normal to the Pore Wall

The variation in the pressure component normal to the surface walls is shown for the liquid (top) and the vapor phases (bottom) in Figure 4, for pore widths, $2a = 17.4\sigma_0$ (left) and $2a = 27.4\sigma_0$ (right), using $\alpha_{sf} = 0.15$ (repelling interactions). The horizontal straight line in all sub-figures is the xx-component extrapolated to the walls. We know there is bulk fluid in the core, as the pressures in V_1 and V_2 are the same, so this component is equal to the bulk pressure. For the other curves in the figure, the tangential components of the pressures (the yy- and zz-components) fall on top of each other. The distances $2a$, b and c between the 4 dashed lines and the pore wall (grey) in the figure represent three fluid volumes, of increasing size (V_1, V_2 and V_{tot}). These are the core volume (width c, volume V_1), the core plus layer 1 (width b, volume V_2) and the whole volume (width $2a$, volume V_{tot}). The distance $2a$ is the width of the channel.

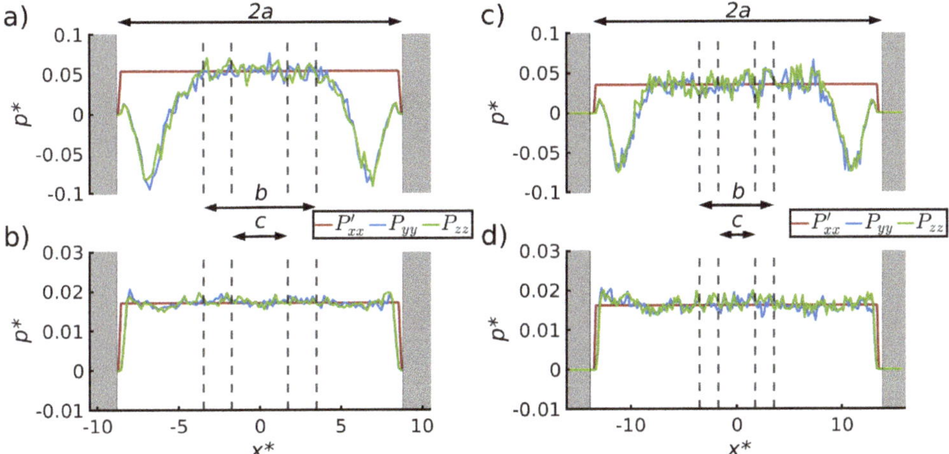

Figure 4. Pressure variation of the three pressure components normal to the surface shown for (**a**) the liquid phase with pore width $17.4\sigma_0$, (**b**) the vapor phase with pore width $17.4\sigma_0$, (**c**) the liquid phase with pore width $27.4\sigma_0$ and (**d**) the vapor phase with pore width $27.4\sigma_0$, for $\alpha_{sf} = 0.15$. The normal component of the pressure, P'_{xx}, is extrapolated from the center of the slit pore.

In the liquid phase (top), the tangential pressure component becomes negative close to the surface. This variation was used to compute the surface tension between solid and liquid from Equation (18). The value was found, within the accuracy of the calculation, to be independent of the distance between the two surface walls, see Table 2.

For the vapor phase, there is no dip in the pressure tensor into the negative regime. In the vapor, not only the extrapolated normal pressure component, P'_{xx}, but also the two tangential components, P_{yy} and P_{zz}, are constant and independent of the distance between the two surfaces. Therefore, the surface tension between the vapor and the wall is negligible.

The pressure in the pore core depends on the distance between the surface walls, i.e., it is larger in the pore with a smaller distance (left) than in the one with a larger distance (right). The bulk fluid pressures are related to the curvature of the liquid-vapor meniscus by Young–Laplace's law, which depends on the slit pore width $2a$.

4.3. Pressure Variation along the Pore

The pressure tensor components were computed for three volumes of increasing size (V_1, V_2 and V_{tot}) using the growing-core methodology. V_1 and V_2 are subvolumes in the center of the REVs and V_{tot} is the volume covering the whole REV, i.e., $V_{\text{tot}} = V^l = V^v$ (see Figures 1 and 3). In Figure 5, the pressure components are plotted as a function of the z-direction. The z-direction is parallel to the pore wall and passes through the vapor and the liquid droplet.

The volumes V_1 and V_2 contain bulk fluid, and are not influenced by the pore wall. The pressure components of the two volumes in the center of the box are equal. It was shown (Table 1) that the pressure was that of a bulk fluid in equilibrium with the pore. Therefore, the liquid bulk pressure is p^l. The pressures are equal to the vapor bulk pressure, p^v, on both sides of the liquid droplet.

The volume V_{tot} includes the fluid and its interface with the pore wall. The pressure component normal to the fluid wall is still equal to the fluid pressure, $P_{xx} = p^f$. The pressure components tangential to the fluid wall in the bulk liquid and bulk vapor, are equal to the integral pressure,

$\hat{p} = \langle P_T \rangle$. The integral pressure is the same in the liquid and vapor phases, in agreement with Equation (20).

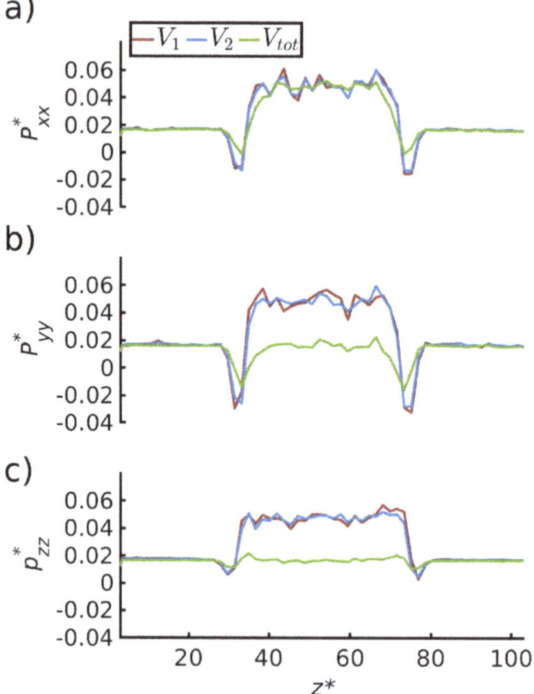

Figure 5. Pressure components (**a**) P_{xx}, (**b**) P_{yy} and (**c**) P_{zz}, as a function of the z- direction for three regions defined using the growing-core methodology. The liquid droplet is in the center of the box, from $z = 30\sigma_0$ to $z = 75\sigma_0$. There is a vapor phase on both sides of the liquid droplet.

4.4. Pressure Differences Across the Liquid-Vapor Interface

Figure 6 (left) shows the difference between the normal components of the bulk pressures in the liquid and vapor. The pressure difference is shown as a function of the inverse half channel width, a, for three values of the interaction parameter α, all values representing wall-fluid repulsive forces. Figure 6 (right) shows the difference between the liquid and vapor for the tangential components of the pressures, i.e., the integral pressure $\hat{p}$ (dashed lines) and the bulk pressure $= p^f$ (solid lines). Results are shown as a function of the inverse half channel width between the two pore walls, $1/a$, and three fluid–solid interactions parameters α. These plots allow us to test Equations (10) and (12).

We first observe that the integral pressure differences for the two fluids are zero, meaning that the integral pressure is constant along the pore, see Figure 6 right bottom. The results support Equation (12). In the bulk phases, the tangential and normal components are the same. This confirms the soundness of the calculation procedures.

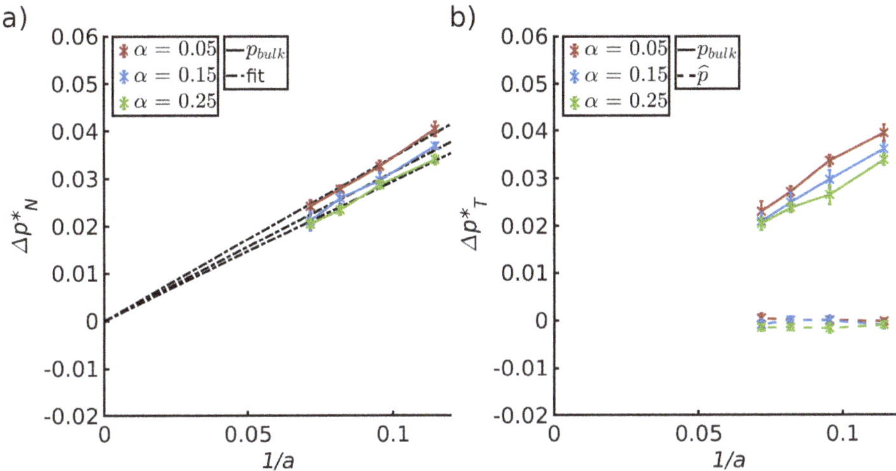

Figure 6. Difference between (**a**) the normal components of the pressure of the bulk liquid and vapor and (**b**) the difference in the tangential component of the pressure of the bulk liquid and vapor, for three different solid-fluid interactions (α) plotted as a function of the inverse half pore width. The bulk pressure, p^f and the integral pressure are represented by solid and dashed lines, respectively.

Both figures show that the difference in bulk pressure components depends on the distance between the two pore walls and the interaction parameter α. This is a consequence of a corresponding decrease of the liquid-vapor curvature. The uncertainties in the results, did not allow a determination of θ. We have therefore not computed the liquid-vapor surface tension, to confirm directly the law Young–Laplace. However, results of the two REVs combined, imply the validity of this law. The liquid–solid and vapor–solid surface tensions, computed with Equation (18), are given in Table 2 for the three α-values. The surface tension changes slightly with α.

Within the accuracy of the calculation, it was possible to estimate the slope of the three data sets. Lines to guide the eye were fitted through zero in the left-hand side figure, following Young–Laplace or the scaling law. The values of the slopes were 0.35/0.31/0.29, for α= 0.05/0.15/0.25, respectively. The slope is equal to $\gamma^{lv} \cos \theta$ which is approximately equal to γ^{ls} since $\gamma^{vs} \approx 0$ (see Equation (14) and Table 2).

The slopes depended on the interaction parameter α. The value increased for decreasing values of α. This is expected: The more repelling the surface is, the higher is the surface tension. The slopes are in agreement with the liquid-solid surface tensions given in Table 2.

Table 2. Surface tensions for the liquid/solid and vapor/solid interfaces. Values are presented for three interaction parameters α in the Lennard-Jones potential used in the computations. The surface tensions were computed via the excess of the tangential pressure tensor (see Equation (18)).

α	γ^{ls}	γ^{vs}
0.05	0.36 ± 0.02	0.0067 ± 0.0026
0.15	0.33 ± 0.01	0.005 ± 0.0035
0.25	0.31 ± 0.01	0.004 ± 0.0045

4.5. The Small System Scaling Law and the Subdivison Potential

Following the definition of Hill, the difference between the differential pressure, p, and the integral pressure, $\hat{p}$, was given by Equation (16), or $\epsilon/V = \gamma^{fs}/(3a)$. The relation is an example of a small system scaling law, cf. Equation (6).

To investigate this further, the differential pressure, p, of the liquid REV minus the integral pressure $\hat{p}$ of the same was plotted in Figure 7 as a function of the inverse half channel width for $\alpha = 0.15$. The difference between p and $\hat{p}$ was proportional to $1/a$, and the slope was as predicted by the liquid/solid surface tension from Table 2 divided by the half channel width, a, and a factor of 3. The line goes through zero, when $1/a$ becomes zero. This is the thermodynamic limit value.

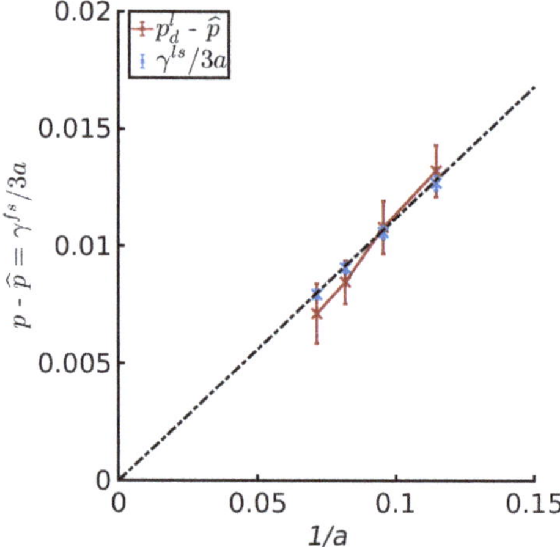

Figure 7. Difference between p and $\hat{p}$ for the liquid REV_2 as a function of the inverse half channel width for $\alpha = 0.15$..

Erdos et al. [10] reported slopes for cylinder pores in addition to slit pores. The ratio of slopes between slit and cylinder pore was 0.52–0.58. From Equations (16) and (17), we obtain the ratio 0.50, when the surface tension is constant. The surface tension is a function of the interaction parameter, as we have seen. The surface tension may also depend on curvature. It follows that results from two pores with different geometries, having the same α, and a surface tension which is independent of curvature, will fall on top of each other.

The scaling law, that relate the integral and differential pressure of a REV, offer a new procedure for calculation of surface tensions. This defined dependence on the inverse characteristic length, may help distinguish between various other types of dependencies (curvature).

5. Conclusions

In this work we have confirmed, using molecular dynamics simulations, that the integral pressure in a nano-confined single fluid in two phases is constant along a pore. This supports and extends earlier investigations of Galteland et al. [18] for a single fluid between spheres arranged in a fcc lattice. The results imply, that Young's and Young–Laplace equations apply, equations which have been validated experimentally numerous times. The integral pressure in mechanical equilibrium can be understood as the average tangential component of the pressure. It is this component that is constant through a pore at equilibrium. The differential pressure at mechanical equilibrium can be understood as the average of the trace of the mechanical pressure tensor divided by 3.

The integral pressure, introduced by Hill, is relevant for descriptions of nano-pores. Using this, we have developed a small system scaling law for nano-pores. Equation (6) is valid for a slit pore and a cylindrical pore. The findings allowed us to speculate on wider applications of the scaling law. One may imagine using the law for more complex geometries, where higher order terms are relevant (line tension contributions). Once the subdivision potential is known, other thermodynamic quantities, like the enthalpy-hat, can also be computed for the confined system [33]. This opens up new ways to study confined systems.

Author Contributions: M.T.R. and O.G. wrote software and carried out the calculations. M.E. did the Monte Carlo simulations. S.K., D.B., M.T.R. and O.G. contributed to the conceptual design and development. The project was supervised by O.A.M., S.K.S., S.K., D.B., and T.J.H.V. All authors contributed to the writing of the manuscript and the scientific discussions. All authors have read and agreed to the published version of the manuscript.

Funding: This research was funded by the Center of Excellence funding scheme of the Research Council of Norway, project no 262644, PoreLab. The work was also sponsored by NWO ExacteWetenschappen (Physical Sciences) for the use of supercomputer facilities, with financial support from the Nederlandse Organisatie voor Wetenschappelijk Onderzoek (Netherlands Organisation for Scientific Research, NWO).

Acknowledgments: The calculation power was granted by The Norwegian Metacenter of Computational Science (NOTUR). Thanks to the Research Council of Norway through its Centres of Excellence funding scheme, project number 262644, PoreLab, and SKS acknowledges funding through the Research Council of Norway and the Young Research Talents Scheme with project number 275754. TJHV acknowledges NWO-CW (Chemical Sciences) for a VICI grant. Discussions with Øivind Wilhelmsen were highly appreciated.

Conflicts of Interest: The authors declare no conflict of interest.

Appendix A

In an isotropic system, the differential pressure p is normally defined by the trace of the pressure tensor [34]:

$$p = \frac{P_{xx} + P_{yy} + P_{zz}}{3} \tag{A1}$$

where P_{ii} is the pressure component in the direction ii. Also the integral pressure can be understood in terms of these pressure components. The integral pressure, $\hat{p}$, in the present slit pore was expressed by Equation (20):

$$\hat{p} = \frac{1}{2a} \int_{-a}^{a} P_T \mathrm{d}x = \langle P_T \rangle = \frac{P_{yy} + P_{zz}}{2} \tag{A2}$$

where P_{yy} and P_{zz} are the pressure components in tangential direction, and x is the direction normal to the surface. The difference between p and $\hat{p}$ can then be written as

$$p - \hat{p} = \frac{P_{xx} + P_{yy} + P_{zz}}{3} - \frac{P_{yy} + P_{zz}}{2} \tag{A3}$$

which results in

$$p - \hat{p} = \frac{P_{xx}}{3} - \frac{P_{yy} + P_{zz}}{6} \tag{A4}$$

By using the definition for the tangential pressure from Equation (A2), we obtain

$$p - \hat{p} = \frac{P_{xx}}{3} - \frac{\langle P_T \rangle}{3} \tag{A5}$$

The normal pressure is equal to the tangential pressure in the thermodynamic limit (the bulk fluid). For the bulk fluid;

$$P_{xx} = \langle P_T \rangle = p^f \tag{A6}$$

These relations mean that there is no difference between p and $\hat{p}$.

With the assumption used in Equation (A2), P_{xx}, is constant in the direction perpendicular to the pore surface, and

$$P_{xx} = p^f \tag{A7}$$

also away from the thermodynamic limit. By combining Equation (A2) with the definition of $\hat{p}$, Equation (11), we obtain

$$\langle P_T \rangle = p^f - \frac{\gamma^{fs}}{a} \tag{A8}$$

By introducing Equations (A7) and (A8) into Equation (A5), we obtain

$$p - \widehat{p} = \frac{p^f}{3} - \frac{p^f}{3} + \frac{\gamma^{fs}}{3a} \tag{A9}$$

which is

$$p - \widehat{p} = \frac{\gamma^{fs}}{3a} \tag{A10}$$

The difference between the differential pressure p and the integral pressure $\widehat{p}$ in this equation is the same as the difference in Equation (16), which was obtained using Hills formalism.

References

1. Keshavarzi, E.; Kamalvand, M. Energy effects on the structure and thermodynamic properties of nanoconfined fluids (a density functional theory study). *J. Phys. Chem. B* **2009**, *113*, 5493–5499. [CrossRef] [PubMed]
2. Braun, E.; Chen, J.J.; Schnell, S.K.; Lin, L.C.; Reimer, J.A.; Smit, B. Nanoporous Materials Can Tune the Critical Point of a Pure Substance. *Angew. Chem.* **2015**, *127*, 14557–14560. [CrossRef]
3. Pharoah, J.; Karan, K.; Sun, W. On effective transport coefficients in PEM fuel cell electrodes: Anisotropy of the porous transport layers. *J. Power Sources* **2006**, *161*, 214–224. [CrossRef]
4. Newman, J.; Tiedemann, W. Porous-electrode theory with battery applications. *AIChE J.* **1975**, *21*, 25–41. [CrossRef]
5. Guo, H.; Wyart, Y.; Perot, J.; Nauleau, F.; Moulin, P. Low-pressure membrane integrity tests for drinking water treatment: A review. *Water Res.* **2010**, *44*, 41–57. [CrossRef]
6. Alklaibi, A.M.; Lior, N. Transport analysis of air-gap membrane distillation. *J. Membr. Sci.* **2005**, *255*, 239–253. [CrossRef]
7. Kuipers, N.; Hanemaaijer, J.H.; Brouwer, H.; van Medevoort, J.; Jansen, A.; Altena, F.; van der Vleuten, P.; Bak, H. Simultaneous production of high-quality water and electrical power from aqueous feedstock's and waste heat by high-pressure membrane distillation. *Desalin. Water Treat.* **2015**, *55*, 2766–2776. [CrossRef]
8. Schnell, S.K.; Vlugt, T.J.H.; Simon, J.M.; Bedeaux, D.; Kjelstrup, S. Thermodynamics of small systems embedded in a reservoir: A detailed analysis of finite size effects. *Mol. Phys.* **2012**, *110*, 1069–1079. [CrossRef]
9. Strøm, B.; Simon, J.M.; Schnell, S.; Kjelstrup, S.; He, J.; Bedeaux, D. Size and shape effects on the thermodynamic properties of nanoscale volumes of water. *Phys. Chem. Chem. Phys.* **2017**, *19*, 9016. [CrossRef]
10. Erdős, M.; Galteland, O.; Bedeaux, D.; Kjelstrup, S.; Moultos, O.A.; Vlugt, T.J.H. Gibbs Ensemble Monte Carlo Simulation of Fluids in Confinement: Relation between the Differential and Integral Pressures. *Nanomaterials* **2020**, *10*, 293. [CrossRef]
11. Hill, T.L. *Thermodynamics of Small Systems*; Dover Publications Inc.: New York, NY, USA, 1994.
12. Dong, X.; Liu, H.; Hou, J.; Wu, K.; Chen, Z. Phase equilibria of confined fluids in nanopores of tight and shale rocks considering the effect of capillary pressure and adsorption film. *Ind. Eng. Chem. Res.* **2016**, *55*, 798–811. [CrossRef]
13. Gubbins, K.E.; Long, Y.; Śliwinska-Bartkowiak, M. Thermodynamics of confined nano-phases. *J. Chem. Thermodyn.* **2014**, *74*, 169–183. [CrossRef]
14. Giovambattista, N.; Rossky, P.J.; Debenedetti, P.G. Phase transitions induced by nanoconfinement in liquid water. *Phys. Rev. Lett.* **2009**, *102*, 050603. [CrossRef] [PubMed]
15. Eslami, H.; Mehdipour, N. Local chemical potential and pressure tensor in inhomogeneous nanoconfined fluids. *J. Chem. Phys.* **2012**, *137*, 144702. [CrossRef] [PubMed]
16. Bennethum, L.S.; Weinstein, T. Three pressures in porous media. *Transp. Porous Med.* **2004**, *54*, 1–34. [CrossRef]
17. Kjelstrup, S.; Bedeaux, D.; Hansen, A.; Hafskjold, B.; Galteland, O. Non-isothermal transport of multi-phase fluids in porous media. Constitutive equations. *Front. Phys.* **2019**, *6*, 150. [CrossRef]
18. Galteland, O.; Bedeaux, D.; Kjelstrup, S.; Hafskjold, B. Pressures inside a nano-porous medium. The case of a single phase fluid. *Front. Phys.* **2019**, *7*, 60. [CrossRef]

19. Bedeaux, D.; Kjelstrup, S.; Schnell, S.K. *Nanothermodynamics. General Theory*; PoreLab, NTNU Grafisk: Trondheim, Norway, 2020.

20. Bedeaux, D.; Kjelstrup, S. Hill's thermodynamics is equivalent with Gibb's thermodynamics for surfaces of constant curvatures. *Chem. Phys. Lett.* **2018**, *707*, 40–43. [CrossRef]

21. Tsai, D. The virial theorem and stress calculation in molecular dynamics. *J. Chem. Phys.* **1979**, *70*, 1375–1382. [CrossRef]

22. Chatterjee, J. Prediction of coupled menisci shapes by Young–Laplace equation and the resultant variability in capillary retention. *J. Colloid Interface Sci.* **2007**, *314*, 199–206. [CrossRef]

23. Gras, J.P.; Delenne, J.Y.; El Youssoufi, M.S. Study of capillary interaction between two grains: A new experimental device with suction control. *Granular Matter* **2013**, *15*, 49–56. [CrossRef]

24. Kirkwood, J.G.; Buff, F.P. The statistical mechanical theory of surface tension. *J. Chem. Phys.* **1949**, *17*, 338–343. [CrossRef]

25. Hafskjold, B.; Travis, K.P.; Hass, A.B.; Hammer, M.; Aasen, A.; Wilhelmsen, Ø. Thermodynamic properties of the 3D Lennard-Jones/spline model. *Mol. Phys.* **2019**, *117*, 3754–3769. [CrossRef]

26. Plimpton, S. Fast parallel algorithms for short-range molecular dynamics. *J. Comput. Phys.* **1995**, *117*, 1–19. [CrossRef]

27. Braga, C.; Travis, K.P. A configurational temperature Nosé-Hoover thermostat. *J. Chem. Phys.* **2005**, *123*, 134101. [CrossRef]

28. Frenkel, D.; Smit, B. *Understanding Molecular Simulation: From Algorithms to Applications*; Elsevier: Amsterdam, The Netherlands, 2001; Volume 1.

29. Hafskjold, B.; Ikeshoji, T. Microscopic pressure tensor for hard-sphere fluids. *Phys. Rev. E* **2002**, *66*, 011203. [CrossRef]

30. Ikeshoji, T.; Hafskjold, B.; Furuholt, H. Molecular-level calculation scheme for pressure in inhomogeneous systems of flat and spherical layers. *Mol. Simul.* **2003**, *29*, 101–109. [CrossRef]

31. Todd, B.; Evans, D.J.; Daivis, P.J. Pressure tensor for inhomogeneous fluids. *Phys. Rev. E* **1995**, *52*, 1627. [CrossRef]

32. Todd, B.D.; Daivis, P.J. *Nonequilibrium Molecular Dynamics: Theory, Algorithms and Applications*; Cambridge University Press: Cambridge, UK, 2017.

33. Skorpa, R.; Simon, J.M.; Bedeaux, D.; Kjelstrup, S. The reaction enthalpy of hydrogen dissociation calculated with the Small System Method from simulation of molecular fluctuations. *Phys. Chem. Chem. Phys.* **2014**, *16*, 19681. [CrossRef]

34. Daivis, P.J.; Evans, D.J. Comparison of constant pressure and constant volume nonequilibrium simulations of sheared model decane. *J. Chem. Phys.* **1994**, *100*, 541–547. [CrossRef]

nanomaterials

Article

Nanothermodynamic Description and Molecular Simulation of a Single-Phase Fluid in a Slit Pore

Olav Galteland *, Dick Bedeaux and Signe Kjelstrup

PoreLab, Department of Chemistry, Norwegian University of Science and Technology, 7491 Trondheim, Norway; dick.bedeaux@ntnu.no (D.B.); signe.kjelstrup@ntnu.no (S.K.)
* Correspondence: olav.galteland@ntnu.no

Abstract: We have described for the first time the thermodynamic state of a highly confined single-phase and single-component fluid in a slit pore using Hill's thermodynamics of small systems. Hill's theory has been named nanothermodynamics. We started by constructing an ensemble of slit pores for controlled temperature, volume, surface area, and chemical potential. We have presented the integral and differential properties according to Hill, and used them to define the disjoining pressure on the new basis. We identified all thermodynamic pressures by their mechanical counterparts in a consistent manner, and have given evidence that the identification holds true using molecular simulations. We computed the entropy and energy densities, and found in agreement with the literature, that the structures at the wall are of an energetic, not entropic nature. We have shown that the subdivision potential is unequal to zero for small wall surface areas. We have showed how Hill's method can be used to find new Maxwell relations of a confined fluid, in addition to a scaling relation, which applies when the walls are far enough apart. By this expansion of nanothermodynamics, we have set the stage for further developments of the thermodynamics of confined fluids, a field that is central in nanotechnology.

Keywords: Hill's thermodynamics of small systems; porous media; integral pressure; molecular simulation

Citation: Galteland, O.; Bedeaux, D.; Kjelstrup, S. Nanothermodynamic Description and Molecular Simulation of a Single-Phase Fluid in a Slit Pore. *Nanomaterials* **2021**, *11*, 165. https://doi.org/10.3390/nano 11010165

Received: 2 December 2020
Accepted: 6 January 2021
Published: 11 January 2021

Publisher's Note: MDPI stays neutral with regard to jurisdictional claims in published maps and institutional affiliations.

1. Introduction

The thermodynamic state of a fluid in confinement is important for the understanding of adsorption to walls, chemical reactions, film formation and transport in porous media [1–6]. The molecular structuring at the walls and the forces between particles and walls are central. The change in thermodynamic properties upon confinement is substantial. This has been known for a long time [1,7]. Derjaguin considered the measurable force that attracts or repels two walls that are close together, and defined from this the disjoining pressure [7]. The disjoining pressure has also been called the solvation pressure [1]. When the walls are far apart, the disjoining pressure vanishes. It is not well known, however, how size and shape, as variables, affect the disjoining pressure or other properties of the confined fluid.

Confinement is considered to be important, for instance, in the context of CO_2 separation and sequestration by metal-organic frameworks [2] or for adsorption in zeolites [3]. The disjoining pressure is of interest when studying aggregation of colloidal particles, suspended or adsorbed [4–6]. It is likely to be important also for film flow on the macroscale [8].

More knowledge of confined fluids on the nanoscale is therefore needed. It may, for instance, help us solve the well-known up-scaling problem in porous media science [9]. The central problem is to understand how to integrate properties on the pore scale to the macroscale where Darcy's law applies. In order to account for shape and size effects, it was recently proposed to use the four Minkowski functionals [10–12]. This simplifies the description of a representative elementary volume (REV). Another procedure using the grand potential for average variables in the REV was also proposed [13,14].

In this work, we want to further examine this procedure [13,14], by looking for a way to describe the confined fluid in a pore. We are looking for a way to deal with size- and shape-dependent variables in a systematic and general manner. Two thermodynamic approaches are common. The approach following Gibbs is most popular [15–19]. However, the method of Hill may offer an attractive alternative [20–25], partly because it may provide an independent check on Gibbs procedure, but also because general geometric scaling relations are obtainable from Hill's method. We will see that this is also the case in the present study.

We will pursue the method of Hill. This starts with the observation that a small system has surface energy comparable to its bulk energy. A consequence is that the properties are not Euler homogeneous. Hill proposed to deal with such systems in an original manner [26]. His idea was to introduce an ensemble of replicas of the small systems, on which standard thermodynamics could be applied. Hill's method has not gained much attention, in spite of a renewed effort to spur interests [27,28].

The long-range aim of this work is to contribute to the effort of finding variables that characterize the confined fluid, for instance in a REV. The grand potential offers one option to describe the pressure and other variables [13,14]. We will pursue this route and study a single-phase and single-component fluid in a slit pore using Hill's method. The so-called integral and differential pressures introduced by Hill are central. Hill did not consider the disjoining pressure, however, it is a small system property and we will see that this concept has benefited from insights of nanothermodynamics. The purpose of this paper is to clarify the use of Hill's nanothermodynamics, by applying the method to a fluid in a slit pore with walls of large surface areas. This is a well-studied case in the literature [1,29], and is well suited to bring out new results on the confined state, the disjoining pressure and other properties. Grand canonical Monte Carlo and molecular dynamics simulation techniques are well suited to investigate thermodynamic relations. We will use these tools to simulate a single-phase fluid in a slit pore in the grand canonical ensemble. A particular advantage of this technique is that the simulations offer a mechanical picture of the system.

Solid–fluid and fluid–fluid interactions are considered here, but not solid–solid interactions. The solid–solid interactions will have a large effect on the thermodynamic state of the system at very small slit pore heights. We do not consider quantum effects that follow from system smallness in this work. We will consider slit pores of height $h/\lambda_B = 6$ to 71, where λ_B is the thermal de Broglie wavelength.

Section 2.1 introduces the reader to Hill's nanothermodynamics. We show in Section 2.2 how this theory can be used to define size- and shape-dependent properties, and new Maxwell relations that follow from these. A definition of the disjoining pressure follows naturally in Section 2.3. In order to be able to verify relations with molecular simulations as a tool, we need to identify the integral pressures and surface tension in terms of the mechanical pressure tensor components. This is done in Section 2.4. In Section 3 we describe the molecular simulations.

We proceed in Section 4 to investigate relations in the theory, and illustrate them with numerical results. We compute the local mechanical and thermodynamic variables according to Hill; i.e., the integral and differential pressures, and the integral and differential surface tensions. The grand potential, or the replica energy, is equal to minus the integral pressure times the volume. The set of thermodynamic variables of the nanothermodynamic framework, in terms of mechanical properties, is found to be self-consistent. We offer concluding remarks in Section 5.

2. Theory

2.1. Hill's Nanothermodynamics

Consider an ensemble of $\mathcal{N}$ slit pores, where each slit pore is filled with a single-phase and single-component fluid. The slit pores do not interact with each other. The j-th slit pore has two parallel plane walls of area Ω_j, separated by a distance h_j. The ensemble of slit pores has the total internal energy U_t, total entropy S_t, total volume $V_t = \sum_{j=1}^{\mathcal{N}} h_j \Omega_j$, total surface area $2\Omega_t = 2\sum_{j=1}^{\mathcal{N}} \Omega_j$, and total number of particles N. The factor of two in the total surface area arises because there are two fluid–solid surfaces of equal area per slit pore. By construction, the ensemble variables $U_t, S_t, V_t, \Omega_t, N,$ and $\mathcal{N}$ are Euler homogeneous functions of the first order in the number $\mathcal{N}$ of slit pores. The total differential of the total internal energy is

$$\mathrm{d}U_t = T\mathrm{d}S_t - p_\perp \mathrm{d}V_t + 2\gamma \mathrm{d}\Omega_t + \mu \mathrm{d}N_t + \varepsilon \mathrm{d}\mathcal{N}. \tag{1}$$

This type of equation for the total internal energy we call the Hill-Gibbs equation [25]. The last term was added by Hill. The partial derivatives of the total internal energy define the temperature T, the normal pressure $p_\perp$, the surface tension γ and the chemical potential μ

$$\begin{aligned}
\left(\frac{\partial U_t}{\partial S_t}\right)_{V_t,\Omega_t,N_t,\mathcal{N}} &= T, &
\left(\frac{\partial U_t}{\partial V_t}\right)_{S_t,\Omega_t,N_t,\mathcal{N}} &= -p_\perp, \\
\left(\frac{\partial U_t}{\partial \Omega_t}\right)_{S_t,V_t,N_t,\mathcal{N}} &= 2\gamma, &
\left(\frac{\partial U_t}{\partial N_t}\right)_{S_t,V_t,\Omega_t,\mathcal{N}} &= \mu.
\end{aligned} \tag{2}$$

The control variables in subscripts are kept constant while taking the derivatives. The volume derivative is taken while keeping the total surface area constant, which implies that the volume is changed by changing the distance between the surfaces $h_j \equiv V_j/\Omega_j$. The surface derivative is taken while keeping the volume constant, which implies that both the pore heights and surface areas are changed in such a way that the change in the total volume is zero.

The new thermodynamic variable ε is the *subdivision potential*. It is

$$\left(\frac{\partial U_t}{\partial \mathcal{N}}\right)_{S_t,V_t,\Omega_t,N_t} = \varepsilon. \tag{3}$$

The subdivision potential is defined here as the increase in the total internal energy as the number of slit pores $\mathcal{N}$ increases while keeping $S_t, V_t, \Omega_t,$ and N_t constant. This definition is different from the definition in a previous article by us [23], where only the entropy, volume and number of particles were kept constant and not the surface area. This led to a different expression for the subdivision potential.

The subdivision potential is the work done on the system when adding a new slit pore while keeping the other control variables constant. The subdivision potential may be positive or negative, depending on whether work is needed or gained by adding new slit pore replicas.

We are aiming to describe the disjoining pressure, see Section 2.3, of an open slit pore when the volume, surface area, temperature and chemical potential are control variables. We will change to this ensemble. This ensemble is useful for describing experiments and simulations. We use the average volume per slit pore $V = V_t/\mathcal{N}$ and the average surface area per slit pore $2\Omega = 2\Omega_t/\mathcal{N}$, rather than the total volume V_t and the total surface area $2\Omega_t$. In order to obtain an appropriate Hill-Gibbs equation for this case, we substitute the total volume with the average volume $V_t = V\mathcal{N}$ and total surface area with the average surface area $2\Omega_t = 2\Omega\mathcal{N}$. The total differentials of the volume and surface area are

$$\mathrm{d}(V\mathcal{N}) = \mathcal{N}\mathrm{d}V + V\mathrm{d}\mathcal{N} \quad \text{and} \quad \mathrm{d}(\Omega\mathcal{N}) = \mathcal{N}\mathrm{d}\Omega + \Omega\mathrm{d}\mathcal{N}. \tag{4}$$

By introducing this into the Hill–Gibbs equation, see Equation (1), we obtain

$$\mathrm{d}U_t = T\mathrm{d}S_t - p_\perp \mathcal{N}\mathrm{d}V + 2\gamma \mathcal{N}\mathrm{d}\Omega + \mu \mathrm{d}N_t + (\varepsilon - p_\perp V + 2\gamma\Omega)\mathrm{d}\mathcal{N}. \tag{5}$$

The parenthesis define the replica energy

$$X(T,V,\Omega,\mu) \equiv \varepsilon - p_\perp V + 2\gamma\Omega. \tag{6}$$

The subdivision potential ε, normal pressure $p_\perp$, and surface tension γ depend on the control variable set T, V, Ω, and μ. The replica energy density will be used to define the disjoining pressure.

The grand partition function covers all microstates available to the slit pore. In order to calculate this partition function one chooses a volume V and a surface area Ω. Both can be varied independently. For large h and Ω only a change in the volume $V = h\Omega$, and not in h and Ω separately, matters. For a small volume, it is necessary to use the volume V and the surface area Ω as independent variables. This is because the same volume change due to a change of the pore height $h = V/\Omega$ or due to a change of the surface area Ω produces different changes in the partition function, and therefore in the thermodynamic variables.

The replica energy was identified by Hill as the grand potential, here equal to minus the integral pressure times the volume [26],

$$X(T, V, \Omega, \mu) = -\hat{p}V = -\hat{p}_\perp V + 2\hat{\gamma}\Omega. \tag{7}$$

In the last equality, we have chosen to identify the integral pressure as the integral normal pressure times the volume minus the integral surface tension times the surface area. We will refer to $\hat{p}_\perp$, $\hat{\gamma}$ and $\hat{p}$ as the integral normal pressure, the integral surface tension and the integral pressure, respectively, while $p_\perp$ and γ are the differential normal pressure and the differential surface tension, respectively. The names integral and differential pressure were coined by Hill to reflect that the differential pressure involves the differential of the integral pressure. We have chosen a control variable set with volume and surface area, such that we do not have a differential pressure but a differential normal pressure $p_\perp$ and differential surface tension γ in its place.

From Equations (6) and (7) it follows that the subdivision potential is

$$\varepsilon = (p_\perp - \hat{p}_\perp)V - 2(\gamma - \hat{\gamma})\Omega. \tag{8}$$

The subdivision potential ε indicates that the integral normal pressure and the integral surface tension may be different from the corresponding differential variables.

2.2. Maxwell Relations for a Slit Pore

Using that the total internal energy is Euler homogeneous of the first order in the number of slit pores, see Equation (5), we integrate it at constant T, V, μ and Ω

$$U_t = TS_t + \mu N_t + X\mathcal{N}. \tag{9}$$

We introduce the average internal energy, entropy and number of particles per slit pore

$$U_t = U\mathcal{N}, \qquad S_t = S\mathcal{N}, \quad \text{and} \quad N_t = N\mathcal{N}. \tag{10}$$

By introducing the average properties into the total internal energy in Equation (9) we obtain the internal energy per slit pore

$$U = TS + \mu N + (\varepsilon - p_\perp V + 2\gamma\Omega) = TS + \mu N + X. \tag{11}$$

Substituting the average properties per slit pore in Equation (10) into the corresponding Hill-Gibbs equation, see Equation (5), and using the internal energy in Equation (11), we obtain the total differential of the internal energy

$$dU = TdS - p_\perp dV + 2\gamma d\Omega + \mu dN. \tag{12}$$

By differentiating the internal energy in Equation (11) and using the total differential of the internal energy in Equation (12), we obtain the total differential of the replica energy

$$dX = -d(\hat{p}_\perp V - 2\hat{\gamma}\Omega) = -d(\hat{p}V) = -SdT - p_\perp dV + 2\gamma d\Omega - Nd\mu. \tag{13}$$

This equation was termed the Hill-Gibbs-Duhem Equation [25], because it reduces to the Gibbs-Duhem equation for a large system. It follows that the partial derivatives of the grand potential is

$$
\begin{aligned}
\left(\frac{\partial(\hat{p}V)}{\partial T}\right)_{V,\Omega,\mu} &= -\left(\frac{\partial X}{\partial T}\right)_{V,\Omega,\mu} = S, \\
\left(\frac{\partial(\hat{p}V)}{\partial V}\right)_{T,\Omega,\mu} &= -\left(\frac{\partial X}{\partial V}\right)_{T,\Omega,\mu} = p_{\perp}, \\
\left(\frac{\partial(\hat{p}V)}{\partial \Omega}\right)_{T,V,\mu} &= -\left(\frac{\partial X}{\partial \Omega}\right)_{T,V,\mu} = -2\gamma, \\
\left(\frac{\partial(\hat{p}V)}{\partial \mu}\right)_{T,V,\Omega} &= -\left(\frac{\partial X}{\partial \mu}\right)_{T,V,\Omega} = N.
\end{aligned}
\tag{14}
$$

Rather than the names replica energy or grand potential, we name from now $X = -\hat{p}V$ by minus the integral pressure times the volume. In Section 2.4 we will define the integral pressure in terms of the average tangential mechanical pressure which can be calculated from molecular simulations. In the simulations, we consider surface areas Ω much larger than the diameter of the fluid particles and heights $h = V/\Omega$ comparable to the diameter of the fluid particles. This implies that $\hat{p}, p_{\perp}, \hat{\gamma}, u \equiv U/V, s \equiv S/V, p_{\perp}, \gamma,$ and $\rho \equiv N/V$ do not depend on Ω, but the variables will depend on the height h and therefore on the volume V.

The volume and surface derivative can be rewritten in terms of derivatives of the slit pore height and surface area. These derivatives are needed to calculate the differential normal pressure and differential surface tension. The differential normal pressure is

$$
p_{\perp} = \left(\frac{\partial(\hat{p}V)}{\partial V}\right)_{T,\Omega,\mu} = \hat{p} + V\left(\frac{\partial \hat{p}}{\partial V}\right)_{T,\Omega,\mu} = \hat{p} + h\left(\frac{\partial \hat{p}}{\partial h}\right)_{T,\Omega,\mu}.
\tag{15}
$$

The differential surface tension is

$$
\gamma = -\frac{V}{2}\left(\frac{\partial \hat{p}}{\partial \Omega}\right)_{T,V,\mu} = \frac{h^2}{2}\left(\frac{\partial \hat{p}}{\partial h}\right)_{T,\Omega,\mu} - \frac{V}{2}\left(\frac{\partial \hat{p}}{\partial \Omega}\right)_{T,h,\mu}.
\tag{16}
$$

Combining the equations for the differential normal pressure and surface tension it follows that the integral pressure is

$$
\hat{p} = p_{\perp} - \frac{2}{h}\gamma - \Omega\left(\frac{\partial \hat{p}}{\partial \Omega}\right)_{T,\mu,h}.
\tag{17}
$$

The last term is the equal to minus the subdivision potential divided by volume, see Equations (7) and (8). The subdivision potential is

$$
\varepsilon = \Omega V\left(\frac{\partial \hat{p}}{\partial \Omega}\right)_{T,\mu,h}.
\tag{18}
$$

This implies that in the grand canonical ensemble with T, V, Ω, μ as control variables, the thermodynamic description of the slit pore with a small height is the same as for the slit pore with a large height. However, it changes when the integral pressure depends on the surface area. This may not be the case for any other set of control variables. In general, the properties of a small system (confined fluid) depend on the set of control variables.

In this work, we deal with large surface areas, such that the integral pressure does not depend on it. As a consequence the subdivision potential is zero. We will find that the integral and differential normal pressure are equal. Using that $\varepsilon = 0$, it follows that the integral and differential surface tensions are also equal for a large surface area, see Equation (8),

$$\hat{p}_\perp = p_\perp \quad \text{and} \quad \hat{\gamma} = \gamma. \tag{19}$$

We will investigate the first identity for large surface areas using molecular simulations.

From Equation (13) we obtain the following Maxwell relations of the differential surface tension,

$$
\begin{aligned}
\left(\frac{\partial \gamma}{\partial T}\right)_{V,\Omega,\mu} &= -\frac{1}{2}\left(\frac{\partial S}{\partial \Omega}\right)_{T,\mu,V} = \frac{h^2}{2}\left(\frac{\partial s}{\partial h}\right)_{T,\mu,\Omega}, \\
\left(\frac{\partial \gamma}{\partial \mu}\right)_{T,V,\Omega} &= -\frac{1}{2}\left(\frac{\partial N}{\partial \Omega}\right)_{T,\mu,V} = \frac{h^2}{2}\left(\frac{\partial \rho}{\partial h}\right)_{T,\mu,\Omega}, \\
\left(\frac{\partial \gamma}{\partial V}\right)_{T,\Omega,\mu} &= -\frac{1}{2}\left(\frac{\partial p_\perp}{\partial \Omega}\right)_{T,\mu,V} = \frac{h}{2\Omega}\left(\frac{\partial p_\perp}{\partial h}\right)_{T,\mu,\Omega}.
\end{aligned}
\tag{20}
$$

In the second identity, we have used that the entropy density, fluid number density and differential normal pressure are independent of the surface area for large surface areas. Other Maxwell relations are possible, see the Hill–Gibbs–Duhem Equation (13). The last equality can be written as

$$\left(\frac{\partial \gamma}{\partial h}\right)_{T,\mu} = \frac{h}{2}\left(\frac{\partial p_\perp}{\partial h}\right)_{T,\mu,\Omega}. \tag{21}$$

The important implication of this expression is that the constant nature of one variable implies the constant nature of the other for large pore heights. With the mechanical description of the integral properties given in Section 2.4 the consistency of all the above thermodynamic relations can be tested.

2.3. The Disjoining Pressure

We define the excess replica energy as the replica energy of a slit pore of height h minus the replica energy of the slit pore where the slit pore height approaches infinity,

$$X^{\text{ex}} = X - \lim_{h\to\infty} X. \tag{22}$$

We denote the thermodynamic variables where the slit pore height approaches infinity with ∞ in superscript,

$$\lim_{h\to\infty} X = X^\infty. \tag{23}$$

The excess replica energy can be written in terms of the excess integral pressure $\hat{p}^{\text{ex}}$, excess integral normal pressure $\hat{p}_\perp^{\text{ex}}$, and excess integral surface tensions $\hat{\gamma}^{\text{ex}}$

$$X^{\text{ex}} = -\hat{p}^{\text{ex}} V = -\hat{p}_\perp^{\text{ex}} V + 2\hat{\gamma}^{\text{ex}}\Omega. \tag{24}$$

We define now the disjoining pressure as the excess normal pressure,

$$\Pi(h) \equiv \hat{p}_\perp^{\text{ex}} \equiv \hat{p}_\perp - p_\perp^\infty. \tag{25}$$

Other possible names are minus the excess replica density, or the excess grand potential density.

This definition of the disjoining pressure is different from the typical definition found in the literature [7,29,30]. Typically the disjoining pressure is defined to be equal to what we in this work call the excess differential normal pressure, where the excess is relative to a bulk fluid. For large surface areas we will show with molecular simulations that $\hat{p}_\perp = p_\perp$. We will furthermore show that the integral normal pressure, as the slit pore height approaches infinite separation, equals the bulk fluid pressure $p_\perp^\infty = p^b$. This shows that our choice is equivalent to the usual definition.

2.4. A Mechanical Description of the Slit Pore

The thermodynamic description of integral and differential pressures and surface tensions has its mechanical equivalent description in terms of components of the mechanical pressure tensor. A recent discussion on this topic clarified the challenge of translating the mechanical pressure tensor into a thermodynamic scalar variable in a meaningful manner [31–33]. We will identify the integral normal pressure, integral surface tension, and integral pressure in such a way that thermodynamic framework is self-consistent. However, we do not claim that this is the only valid choice of thermodynamic pressures and tensions in terms of the mechanical pressure tensor.

The mechanical pressure tensor of a heterogeneous system is ambiguously defined. This has been known for a long time, at least since the work by Irving and Kirkwood in the 1940s. It was shown by Schofield and Henderson [34] that the ambiguity is due to the arbitrary choice of the integration contour C_{ij}, which is needed to calculate the configurational contribution to the pressure tensor. The local mechanical pressure tensor is calculated in a subvolume V_l as a sum of the kinetic and the configurational contributions,

$$P_{\alpha\beta}(x) = P_{\alpha\beta}^{k} + P_{\alpha\beta}^{c}. \tag{26}$$

Upper case P is used to denote mechanical pressure tensor components in order to distinguish them from the thermodynamic pressures, which are denoted by the lower case p. The kinetic contribution is the ideal contribution to the mechanical pressure and is calculated as

$$P_{\alpha\beta}^{k} = \frac{1}{V_l} \left\langle \sum_{i \in V_l} m_i v_{i,\alpha} v_{i,\beta} \right\rangle. \tag{27}$$

The sum with subscript $i \in V_l$ represents a sum over all particles in the subvolume V_l. The particle mass is m_i and $v_{i,\alpha}$ is the velocity in the α-direction. The solid walls do not have a velocity and consequently do not directly contribute to the kinetic pressure. The brackets $\langle \dots \rangle$ represent ensemble average. The configurational contribution is the non-ideal contribution to the mechanical pressure and is calculated as

$$P_{\alpha\beta}^{c} = -\frac{1}{2V_l} \left\langle \sum_{i=1}^{N} \sum_{\substack{j=1 \\ j \neq i}}^{N} f_{ij,\alpha} \int_{C_{ij} \in V_l} \mathrm{d}l_{\beta} \right\rangle. \tag{28}$$

The sum represents a sum over all particle pairs. The α-component of the force vector acting on particle i due to particle j is $f_{ij,\alpha}$. The fluid–fluid and fluid–solid interactions contribute to the configurational pressure, the solid–solid interaction is zero and does not contribute to the pressure. The line integral is the β-component of the part of the contour C_{ij} contained in the subvolume V_l.

The contour C_{ij} is the source of the ambiguity of the mechanical pressure tensor, it can be any continuous line from the centers of particles i to j. The Harasima [35] and the Irving–Kirkwood [36] contours are two common choices for C_{ij}. The Harasima contour is defined as two continuous line segments, a line from the center of particle i parallel to the surface and a line normal to the surface to the center of particle j. The Irving–Kirkwood is the straight line from particle i to j. For flat surfaces they are equal. However, for spherical surfaces, the Harasima contour does not obey momentum balance [37]. There are cases where the Harasima contour is more useful than the Irving–Kirkwood contour [38]. In this work, we will use the equations by Ikeshoji et al. [39] with the Irving–Kirkwoood contour to calculate the mechanical pressure tensor.

We will only consider surface areas Ω much larger in both directions than the fluid particle diameter. This implies the mechanical pressure tensor does not depend on Ω. It does, however, depend on the height $h \equiv V/\Omega$ and therefore on the volume V.

Using the translational symmetry of the slit pore in the y- and z-direction, where the x-direction is normal to the solid surface, the equilibrium mechanical pressure tensor in the slit pore has the form

$$\mathbf{P}(x;h) = P_\perp(h)\mathbf{e}_x\mathbf{e}_x + P_\parallel(x;h)(\mathbf{e}_y\mathbf{e}_y + \mathbf{e}_z\mathbf{e}_z), \tag{29}$$

where $\mathbf{e}_x$, $\mathbf{e}_y$ and $\mathbf{e}_z$ are the unit vectors in the x-, y- and z-directions. The normal pressure tensor component is equal to the xx-component, and tangential pressure tensor component is the average of the yy- and zz-components,

$$P_\perp(h) = P_{xx} \quad \text{and} \quad P_\parallel(x,h) = \frac{1}{2}(P_{yy} + P_{zz}). \tag{30}$$

Mechanical equilibrium requires that the tangential pressure is independent of the y- and z-coordinates, but depends on the x-coordinate. The normal pressure is independent of all spatial coordinates.

We identify the thermodynamic integral normal pressure in terms of the volume integral of the normal mechanical pressure divided by the volume. However, since the normal mechanical pressure and the area are constant everywhere, this simplifies to

$$\hat{p}_\perp(h) \equiv \frac{1}{h}\int_0^h P_\perp(h)x = P_\perp(h). \tag{31}$$

The integral normal pressure in a large pore is equal to the pressure in a bulk phase in equilibrium with the pore. The difference in the integral normal pressure in a liquid and vapor phase in a slit pore is described the Young–Laplace equation, while the integral pressure is the same in both the liquid and vapor phase [23]. An alternative route to the integral normal pressure is via the local fluid number density profile [40]

$$P_\perp(h) = \int_0^h f_{fs}(x)\rho(x,h)x, \tag{32}$$

where $f_{fs}(x)$ is the fluid–solid force and $\rho(x,h)$ is the local fluid number density. The fluid density and fluid–solid forces are uniquely defined, and do not have the inherent problem that the mechanical pressure tensor has. We can use this equation to validate our method of calculating the integral normal pressure.

We identify further the integral surface tension as the integral of the normal minus tangential pressure tensor components,

$$\hat{\gamma}(h) \equiv \frac{1}{2}\int_0^h \Big(P_\perp(h) - P_\parallel(x;h)\Big)\mathrm{d}x, \tag{33}$$

where the factor half is due to the fact that there are two fluid–solid surfaces. It follows from Equation (7) together with Equations (31) and (33) that the integral pressure is,

$$\hat{p}(h) = \frac{1}{h}\int_0^h P_\parallel(x;h)\mathrm{d}x. \tag{34}$$

As shown by Harasima [35] and Schofield and Henderson [34] a sufficiently large volume integral of the mechanical pressure tensor components does not depend on the choice of the integral contour C_{ij}. We have identified the integral normal pressure, integral surface tension and integral pressure in terms of the mechanical pressure tensor. The local mechanical pressure tensor has an inherent problem, specifically that the contour C_{ij} can be any continuous line from i to j. However, the thermodynamic variables do not have this inherent problem. This is because we integrate the local mechanical pressure tensor across the whole volume V. This volume integral includes all interactions.

The internal energy can be calculated as the sum of the kinetic and potential energy,

$$U = E_k + E_p. \tag{35}$$

By dividing the internal energy in Equation (11) by the volume we obtain the internal energy density $u = U/V$. We also use the entropy density $s = S/V$, and fluid number density $\rho = N/V$. By rearranging the equation we obtain the entropy density as

$$s = \frac{1}{T}(u + \hat{p} - \mu\rho), \tag{36}$$

where the internal energy density, integral pressure and fluid number density are known. The volume, chemical potential and temperature are imposed on the system.

3. Simulation Details

The thermodynamic state of slit pores of varying heights h was investigated by using grand-canonical Monte Carlo (GCMC) [41] in combination with molecular dynamic (MD) simulations with the Nosé–Hoover thermostat [42]. This produced the grand canonical ensemble, i.e., constant chemical potential, temperature, volume and surface area. The GCMC method inserted and deleted fluid particles to and from the simulation box from an imaginary fluid particle reservoir at the same temperature and chemical potential. This controlled the chemical potential of the fluid in the slit pore. The MD procedure updated the positions and velocities of the fluid particles and controlled the temperature with the Nosé–Hoover thermostat.

The simulations were carried out using LAMMPS [43]. The local mechanical pressure tensor was calculated by post-processing the particle trajectories with in-house software (available at D.O.I. 10.5281/zenodo.4405267). The chemical potential and temperature were kept constant at $\mu^* = 1$ and $T^* = 2$. The critical temperature of the Lennard–Jones/spline fluid is $T_c^* = 0.885$ [44]. All units in this work are in reduced Lennard–Jones units, see Table 1 for a definition.

Table 1. The reduced units are denoted with an asterisk in superscript, for example T^*. The variables are reduced using the molecular diameter σ, potential well depth ϵ, fluid particle mass m and Boltzmann constant k_B.

Description	Definition
Energy	$E^* = E/\epsilon$
Entropy	$S^* = S/k_B$
Temperature	$T^* = Tk_B/\epsilon$
Distance	$x^* = x/\sigma$
Pressure	$p^* = p\sigma^3/\epsilon$
Chemical potential	$\mu^* = \mu/\epsilon$

The simulation box was a rectangular cuboid of side lengths $L_x, L_y = L_z$. The side lengths $L_y = L_z$ were chosen such that the surface area was large. Large in this context indicates large enough for $\hat{p}$, $\hat{p}_\perp$, $\hat{\gamma}$, $p_\perp$, and γ to be independent of the surface area $\Omega = L_y L_z$. The simulation box size was decided such that the average number of fluid particles was approximately 2×10^4. The simulation box was periodic in the y- and z-directions, and non-periodic in the x-direction. This implies that the particles did not interact across the simulation box boundary in the x-direction.

The fluid–fluid and fluid–solid interaction was modeled with the Lennard-Jones/spline potential [44]. The fluid–fluid and fluid–solid interactions were equal. The potential energy of a fluid–fluid or fluid–solid pair separated by a distance r was

$$u^{LJ/s}(r) = \begin{cases} 4\epsilon\left[\left(\frac{\sigma}{r}\right)^{12} - \left(\frac{\sigma}{r}\right)^6\right] & \text{if } r < r_s \\ a(r - r_c)^2 + b(r - r_c)^3 & \text{if } r_s < r < r_c \\ 0 & \text{else,} \end{cases} \tag{37}$$

where r_s, a, b and r_c were chosen such that the potential energy and the force were continuous at the inflection point $r = r_s$ and the cut-off $r = r_c$. The solid walls were placed at the simulation box boundaries $x = -L_x/2$ and $x = L_x/2$. The distance between the fluid and solids were $r = |x_f - L_x/2|$ and $r = |x_f + L_x/2|$, where x_f is the x-position of the fluid particle.

The dividing surfaces of the fluid–solid surfaces were chosen to be at $x = -L_x/2$ and $x = L_x/2$. The slit pore height was consequently determined to be $h = L_x$. Other choices of the dividing surface are possible, for example, the Gibbs dividing surface or the surface of tension. When $L_x < 2r_c$ the fluid particle can interact with both solid walls. See Figure 1 for an visualization of the simulation box for the case $L_x = 4\sigma$.

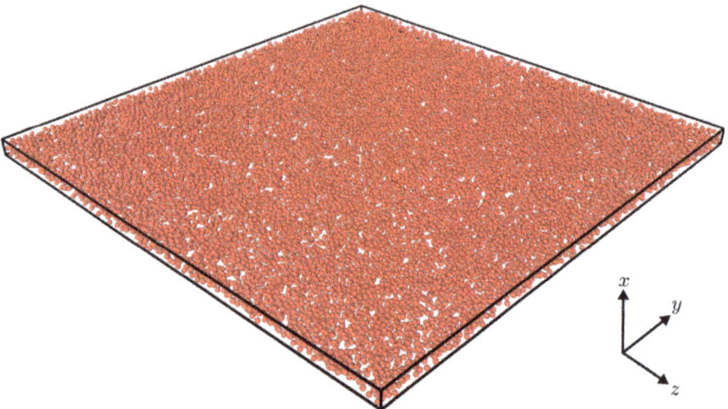

Figure 1. Visualization of the fluid particles in a slit pore of height $L_x = 4\sigma$, chemical potential $\mu^* = 1$, and temperature $T^* = 2$. The fluid particles are rendered in red, and their diameter is rendered at σ. The solid is not rendered. The solid lines illustrates the edges of the simulation box. The simulation was rendered with Open Visualization Tool (OVITO) [45].

The mechanical pressure tensor was calculated in thin rectangular cuboids, called layers l, of side lengths $\Delta x, L_y, L_z$. The thickness of the layers was $\Delta x = 0.005\sigma$ and the number of layers was $n_l = h/\Delta x$. The diagonal components of the mechanical pressure tensor was calculated using Equations (27)–(29).

The kinetic energy was calculated as the sum of the kinetic energy for each fluid particle and the potential energy was calculated as the sum of the potential energy of each fluid–fluid and fluid–solid pairs,

$$E_k = \frac{1}{2}\sum_{i=1}^{N} m_i(\mathbf{v}_i \cdot \mathbf{v}_i) \quad \text{and} \quad E_p = \sum_{i=1}^{N}\sum_{j>i}^{N} u^{LJ/s}(r_{ij}). \tag{38}$$

The sums of the kinetic and potential energies were used to calculate the internal energy and entropy densities.

4. Results and Discussion

The results are presented in Figures 2–10 and discussed in that order before general remarks are offered.

The normal mechanical pressure $P_\perp$ is presented in Figure 2a. It does not depend on the position x, but it depends strongly on the slit pore height h. The figure shows a straight line of various lengths for each of the three heights, which reflect the slit pore height h. The integral normal pressure was identified as this component, $\hat{p}_\perp = P_\perp$. We see that it is always constant, as demanded by Equation (31). For slit pore heights $h > 7\sigma$ we find that the normal mechanical pressure is equal to the bulk pressure. At $h = 2.04\sigma$ the normal mechanical pressure is at a global maximum and at $h = 2.59\sigma$ it has a local minimum for the given temperature and chemical potential. The normal mechanical pressures divided by the bulk pressure for the two cases are $P_\perp/p^b = 2.5135 \pm 0.0008$ and $P_\perp/p^b = 0.5539 \pm 0.0002$, respectively. The global minimum, which is zero, is at $h < 1.8\sigma$ when no fluid particles fit in the slit pore. This is because solid–solid interactions and quantum effects are not considered in this work.

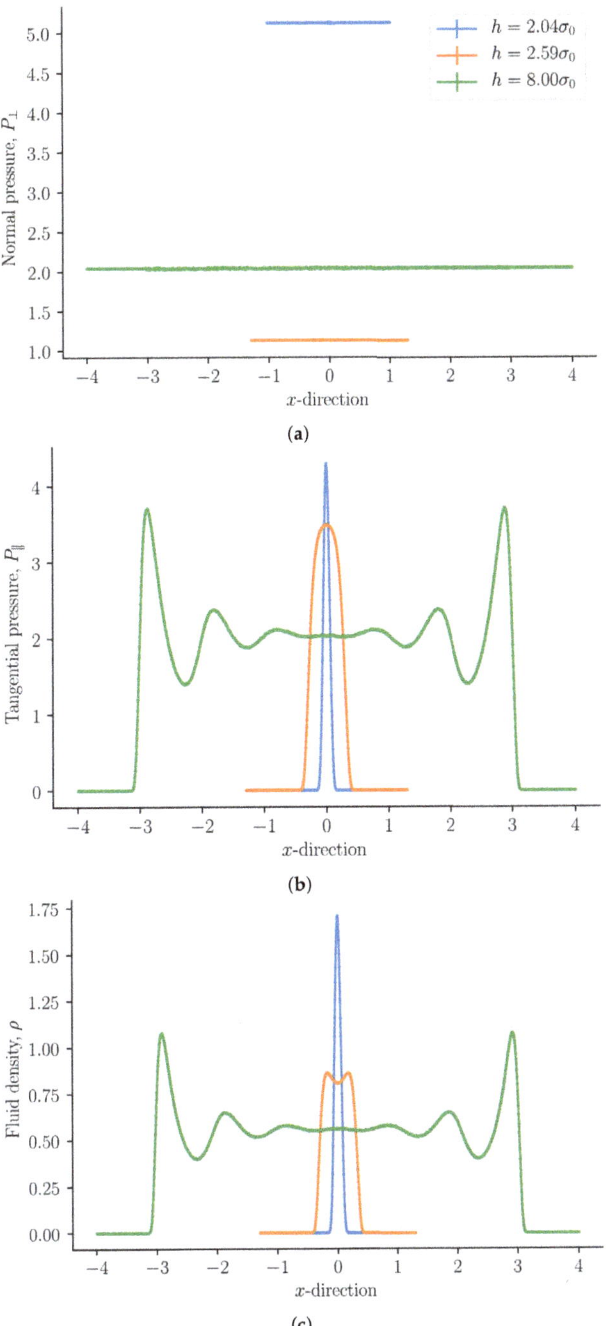

Figure 2. (**a**) Normal mechanical pressure, (**b**) tangential mechanical pressure, and (**c**) fluid number density ρ as a function of th e x-direction for slit pore heights $h = 2.04\sigma$, 2.59σ, and 8σ.

The tangential mechanical pressure $P_\parallel$, illustrated in Figure 2b, depends in contrast on the position x as well as on the slit pore height h. The integral pressure is the average of $P_\parallel$, see Equation (34). The tangential mechanical pressure follows the trend of the fluid number density, compare Figures 2b,c. For pore sizes $h > 7\sigma$ the tangential mechanical pressure is constant and equal to the bulk pressure p^b in the center of the pore. This indicates that the pore is large enough to accommodate bulk liquid in the center. The fluid is highly structured close to the fluid–solid surface [1]. As the slit pore height is decreased, fluid structures on the two sides overlap. When regions of structured fluids overlap, repulsive and attractive forces between the surfaces appear, and the disjoining pressure becomes non-zero.

The fluid number density $\rho = N/V$ is presented in Figure 3 as a function of the slit pore height h. The bulk fluid number density ρ^b in equilibrium with the slit pore is shown as a dashed line. The fluid number density converges to the bulk value as the slit pore height approaches infinity. The volume V depends on the choice of the fluid–solid dividing surfaces. We have chosen the dividing surfaces to be at $x = -L_x/2$ and $x = L_x/2$. The choice of the dividing surface determines how rapidly the slit pore values converge to the bulk values. A dividing surface closer to the fluid phase will reduce the volume and consequently the slit pore values will converge faster to the bulk values. Other choices of the dividing surface are possible.

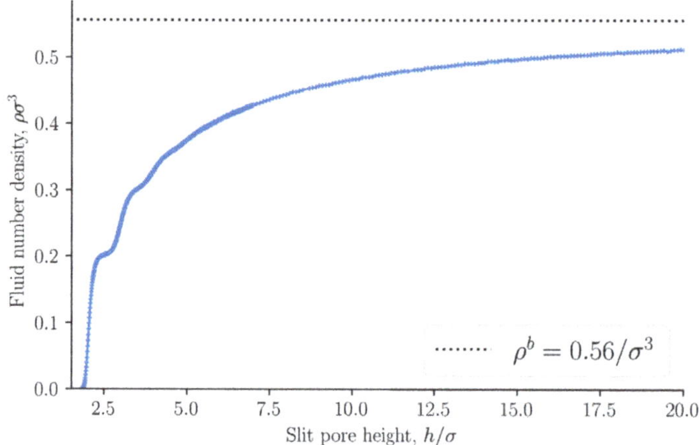

Figure 3. Fluid number density $\rho = N/V$ as a function of slit pore height h. The bulk fluid number density is shown as a dashed line.

The entropy density is presented in Figure 4 as a function of the slit pore height h. The entropy density is a monotonically increasing function of the height h. This confirms the observation by Israelachvili [1] that the origin of the oscillations of the disjoining pressure as a function of the height is not entropic. As a further confirmation of this point, we find that the internal energy density oscillates with a period equal to the particle diameter, see Figure 5. The oscillating forces or pressures are thus of energetic origin. The bulk entropy and internal energy densities are shown as dashed lines. The entropy and internal energy densities of the slit pore converge to the bulk values as the slit pore height is increased.

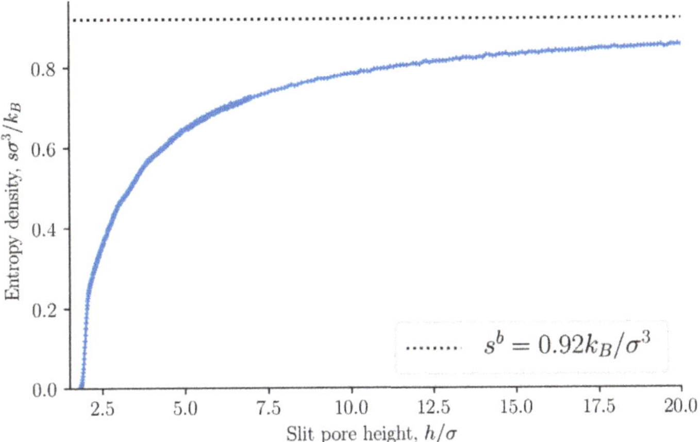

Figure 4. Entropy density $s = S/V$ as a function of slit pore height h. See Equation (36). The dashed line shows the entropy density of the bulk s^b.

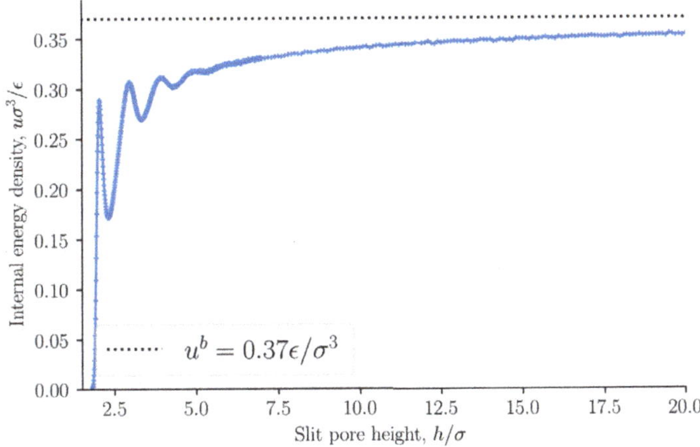

Figure 5. Internal energy density $u = U/V$ as a function of slit pore height h. See Equation (35). The dashed line shows the internal energy density of the bulk u^b.

The integral pressure is equal to the volume average tangential mechanical pressure $\hat{p} = h^{-1} \int_0^h P_\parallel \mathrm{d}x$. It is of special interest because it is equal to minus the grand potential divided by volume $\hat{p} = -X/V$, or the replica energy density. The grand potential is the starting point for the definition of the average thermodynamic properties of the REV [13]. The integral pressure is presented in Figure 6 as a function of the slit pore height h. The integral pressure converges to the bulk pressure p^b as the slit pore height approaches infinity, as expected. The bulk pressure is shown as a dashed line.

In previous works [13,14,21] we argued that the gradient of the integral pressure is the driving force for mass flux. In another work [23] we found the integral pressure of a two-phase system in a slit pore to be equal in the liquid and vapor in equilibrium. The identification of the integral pressure in this work is consistent with this interpretation. The gradient of the integral pressure is the driving force of the mass flux. For fluid flows tangential to the slit pore surfaces it is the gradient in the tangential mechanical pressure tensor component that gives the driving force when the system is out of equilibrium. In this work we identify the integral pressure as the average of the tangential mechanical pressure tensor components.

As stated in Section 2.2, the integral and differential normal pressures and integral and differential surface tensions are expected to be equal when the surface area is large. If this is correct, the integral pressure can be computed as

$$\hat{p}(h) = \frac{1}{h}\int_{h_0}^{h}\hat{p}_\perp h' \quad = \int_{h_0}^{h}\frac{2\hat{\gamma}}{h'^2}h'. \tag{39}$$

The lower integration limit is $h_0 = 1.8\sigma$, at which point the integral pressure is in good approximation zero. The integral pressure computed from Equation (39) is shown in Figure 6. The curves are identical. This implies that the integral and differential normal pressures are equal. As we have already shown that the subdivision potential is zero $\varepsilon = 0$ for large surface areas with this set of control variables, it follows that the integral and differential surface tensions are also equal. We will from now on refer to the integral normal pressure and integral surface tensions as the normal pressure and surface tension.

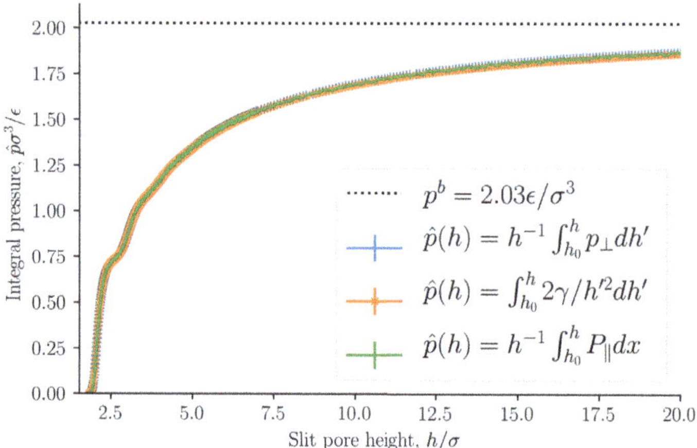

Figure 6. Integral pressure $\hat{p}$ as a function of slit pore height h. The integral pressure is computed as the average tangential mechanical pressure, see Equation (34), and from the normal pressure and surface tension, see Equation (39).

The normal pressure was identified as the normal mechanical pressure $p_\perp = P_\perp$ in Equation (31). It is presented in Figure 7 as a function of the slit pore height h. The normal pressure was also calculated from the local fluid density and the fluid–solid force using Equation (32). The two methods of calculating the normal pressure agree, which indicates that we have calculated the mechanical pressure tensor correctly.

The normal pressure oscillates with a period equal to the fluid particle diameter at small heights h. The oscillations decay as the height h increases. Such oscillations have been observed in experiments and are well known, see for example Israelachvili [1]. The oscillations are caused by the structuring of the fluid particles between the surfaces, and by the fact that the fluid particles all have the same diameter. As the height is increased above $h > 7\sigma$ the oscillations vanish and the normal pressure is constant and equal to the bulk pressure. The bulk pressure is shown as a dashed line in the figure. At heights $h > 7\sigma$ the fluid structuring near the walls do not overlap. At lower densities, smaller oscillations are expected with a faster decay. The normal pressure shows a similar trend to previous works [15,16,40]. At very small heights, the solid–solid interaction will dominate and completely overshadow the fluid-fluid and fluid–solid interactions presented here. We have not included any solid–solid interaction in this work, and as a consequence the normal pressure approaches zero because there is no room for any fluid particles to enter the slit pore.

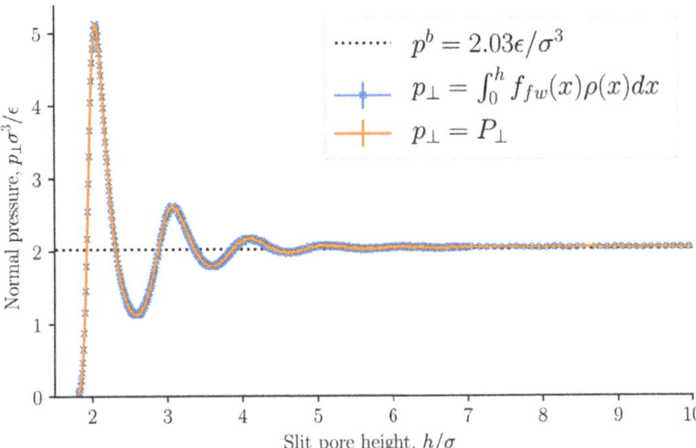

Figure 7. Normal pressure $p_\perp = P_\perp$ as a function of slit pore height h. It is computed as the normal mechanical pressure tensor component, see Equation (31), and as the integral of the fluid–solid force times the local fluid density, see Equation (32).

When $p_\perp = \hat{p}_\perp$ and $\varepsilon = 0$, it follows from Equation (8) that the integral and differential surface tensions are equal $\gamma = \hat{\gamma}$. The surface tension as a function of the slit pore height is presented in Figure 8, see Equation (33). The surface tension at infinite separation γ^∞ is computed as the average surface tension of the slit pore, with height $h > 10\sigma$, at which point the surface tension is independent of h.

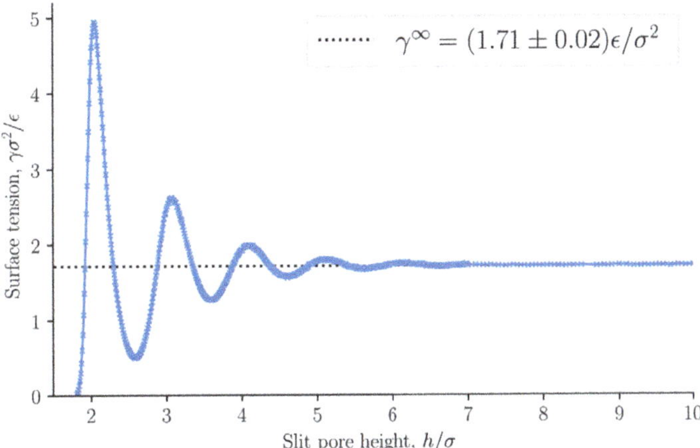

Figure 8. Surface tension γ as a function of slit pore height h, see Equation (33). The dashed line shows the surface tension at infinite separation γ^∞.

The disjoining pressure Π was computed from Equation (25), and is shown as a function of the slit pore height h in Figure 9. The disjoining pressure was here defined to be equal to the integral normal pressure minus the normal pressure at infinite separation. Because the integral and the differential normal pressures are the same and because the normal pressure at infinite separation is equal to the bulk pressure, in this case, the definition contain the commonly used definition of the disjoining pressure [7,29]. The normal pressure is constant for slit pore heights $h > 7\sigma$. Consequently, the normal pressure at infinite separation can be calculated as the normal pressure when $h > 7\sigma$. The normal pressure at infinite separation is equal to the pressure in a bulk fluid with the same temperature and chemical potential.

For the present case, we can claim that our definition of the disjoining pressure is equivalent to the common definition. Our definition is general, as it also covers the cases where the integral and the differential normal pressures are unequal. Examples where this is the case are given below.

Figure 10 shows how the normal pressure minus the integral pressure scales with inverse slit pore height. It is equal to the scaling law presented in a previous work [23]. The slope of this ideal curve is equal to two times the surface tension. When the inverse slit pore height approaches zero, i.e., when the walls are far apart, the normal pressure and integral pressures are equal as predicted. For pores that are so small that no bulk fluid can form in the center, the structuring at the walls starts to overlap, and a fluctuating difference is seen in the difference of the normal and integral pressures. In the region of the straight line, we have a scaling law, that relates states of different heights. At heights smaller than approximately $h < 5\sigma$ the scaling law breaks down, the difference of the two pressures starts to oscillate. There are positive and negative deviations from the law.

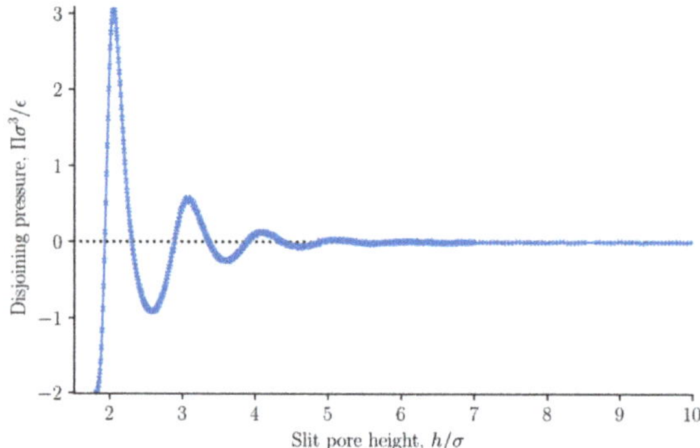

Figure 9. Disjoining pressure as a function the slit pore height h, see Equation (25).

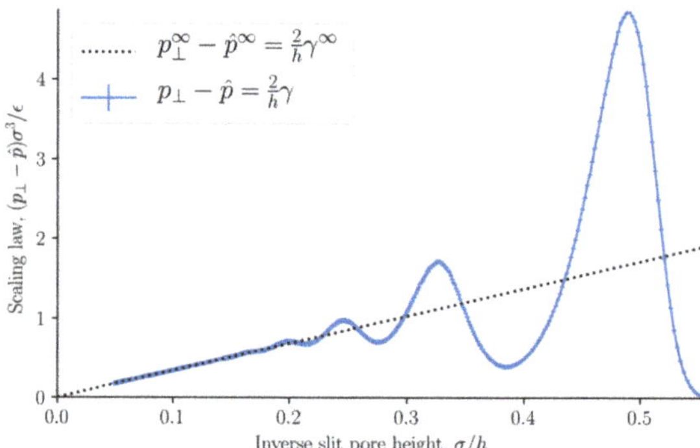

Figure 10. The scaling of normal pressure minus integral pressure as a function of the inverse slit pore height h.

In our earlier work [23], we studied liquid-vapor coexistence in a slit pore. In that work the surface area was not a control variable, and as a consequence the subdivision potential was found to be equal to two times the surface tension divided by the slit pore height, see Figure 10. Another reasonable set of control variables is the height h instead of the volume V. For a control variable set consisting of temperature, height, surface area and chemical potential, the differential normal pressure is equal to

$$p_\perp = \frac{1}{\Omega}\left(\frac{\partial(\hat{p}V)}{\partial h}\right)_{T,\Omega,\mu} = \hat{p} + h\left(\frac{\partial \hat{p}}{\partial h}\right)_{T,\Omega,\mu}. \tag{40}$$

The differential surface tensions is

$$\gamma = -\frac{1}{2}\left(\frac{\partial(\hat{p}V)}{\partial \Omega}\right)_{T,\mu,h} = -\frac{\hat{p}h}{2} - \frac{V}{2}\left(\frac{\partial \hat{p}}{\partial \Omega}\right)_{T,\mu,h}. \tag{41}$$

For this set of control variables, the height h is kept constant instead of the volume V. The subdivision potential is accordingly

$$\begin{aligned}
\varepsilon &= -\hat{p}V + p_\perp V + 2\gamma\Omega \\
&= \hat{p}V + hV\left(\frac{\partial \hat{p}}{\partial h}\right)_{T,\Omega,\mu} + \Omega V\left(\frac{\partial \hat{p}}{\partial \Omega}\right)_{T,h,\mu}.
\end{aligned} \tag{42}$$

These relations also help us characterize the smallness of the slit pore with the large walls. The subdivision potential, introduced by Hill as a measure of smallness, deviates from zero in the last relation, also when the integral pressure does not depend on the size of the area, with height h and surface area Ω as control variables. A dependency on the area is relevant when adsorption takes place on small spheres [24]. A slit pore with large walls may be expected to be small for small heights h, since the confined fluid is not bulk-like. However, we have seen that when we use volume V and surface area Ω as control variables, the subdivision potential is zero for large surface areas. A zero subdivision potential means that the system also can be described perfectly using regular thermodynamics [18]. It is nevertheless meaningful to define a non-zero integral pressure, because the integral pressure enters the grand potential. It, therefore, determines the thermodynamic properties of a REV. Away from equilibrium, it will create a driving force.

The grand potential or minus the integral pressure times the volume are equal to the replica energy. The replica energy is not zero in the present case. Clearly, we have here an example where smallness is brought out in Hill's terms through the replica potential.

5. Conclusions

We have developed a nanothermodynamic description based on the ideas of Hill to describe single-phase and single-component fluids in slit pores in a new manner. As environmental control variables we chose the chemical potential, temperature, volume and surface area. We have seen that the outcome varies with the set chosen, but the procedure can be used for complex geometries and different sets.

Following Hill, we introduced the subdivision potential. It is non-zero only when the integral pressure depends on the surface area Ω. For large surface areas we have shown that the subdivision potential is zero, and that $p_\perp = \hat{p}_\perp$ and $\gamma = \hat{\gamma}$. In this sense, we have found that nanothermodynamics is equivalent to the usual thermodynamic description for all heights. The replica energy, and therefore the integral pressure, were shown to be non-zero. This allowed us to identify a scaling law, which confirms earlier results [22,23]. By choosing height h rather than volume V among the control variables, a non-zero subdivision potential appears.

We have identified the thermodynamic properties by their mechanical counterparts in a consistent manner. The integral pressure, which is equal to minus the grand potential divided by volume, can be understood as the average tangential mechanical pressure. The normal pressure is the normal mechanical pressure, and the surface tension is the integral of the normal minus the tangential mechanical pressure. The entropy and internal energy densities vary with the slit pore height, confirming the observations of Israelachvili [1]. The entropy density increases monotonically with increasing height, while the energy density oscillates. This confirms that the disjoining pressure is not of entropic origin, it is of energetic origin [1].

By these investigations of the nanothermodynamic theory and the subsequent simulations, we hope to have expanded on the knowledge on Hill's method, making it more available for further studies, for instance of flow and reactions in porous media.

Author Contributions: O.G. contributed to investigation, methodology, software and visualization. D.B. and S.K. contributed to supervision. All authors contributed to conceptualization, formal analysis, writing original draft, reviewing and editing. All authors have read and agreed to the published version of the manuscript.

Funding: We are grateful to the Research Council of Norway for funding through its Center of Excellence funding scheme, project number 262644, PoreLab.

Institutional Review Board Statement: Not applicable.

Data Availability Statement: The data set is available at D.O.I. 10.5281/zenodo.4405271.

Acknowledgments: The computations were performed on resources provided by UNINETT Sigma2—the National Infrastructure for High Performance Computing and Data Storage in Norway.

Conflicts of Interest: The authors declare no conflict of interest.

References

1. Israelachvili, J.N. *Intermolecular and Surface Forces*; Academic Press: Cambridge, MA, USA, 1985.
2. McDonald, T.M.; Mason, J.A.; Kong, X.; Bloch, E.D.; Gygi, D.; Dani, A.; Crocella, V.; Giordanino, F.; Odoh, S.O.; Drisdell, W.S.; et al. Cooperative insertion of CO_2 in diamine-appended metal-organic frameworks. *Nature* **2015**, *519*, 303–308. [CrossRef] [PubMed]
3. Vlugt, T.; Krishna, R.; Smit, B. Molecular simulations of adsorption isotherms for linear and branched alkanes and their mixtures in silicalite. *J. Phys. Chem. B* **1999**, *103*, 1102–1118. [CrossRef]
4. Bresme, F.; Oettel, M. Nanoparticles at fluid interfaces. *J. Phys. Condens. Matter* **2007**, *19*, 413101. [CrossRef] [PubMed]
5. Bresme, F.; Lehle, H.; Oettel, M. Solvent-mediated interactions between nanoparticles at fluid interfaces. *J. Chem. Phys.* **2009**, *130*, 214711. [CrossRef]
6. Galteland, O.; Bresme, F.; Hafskjold, B. Solvent-Mediated Forces between Ellipsoidal Nanoparticles Adsorbed at Liquid–Vapor Interfaces. *Langmuir* **2020**, *36*, 48. [CrossRef]
7. Derjaguin, B. Untersuchungen über die Reibung und Adhäsion, IV. *Kolloid Z.* **1934**, *69*, 155–164. [CrossRef]
8. Moura, M.; Flekkøy, E.G.; Måløy, K.J.; Schäfer, G.; Toussaint, R. Connectivity enhancement due to film flow in porous media. *Phys. Rev. Fluids* **2019**, *4*, 094102. [CrossRef]
9. Das, D.; Hassanizadeh, S. *Upscaling Multiphase Flow in Porous Media*; Springer: Berlin, Germany, 2005.
10. Khanamiri, H.H.; Berg, C.F.; Slotte, P.A.; Schlüter, S.; Torsæter, O. Description of Free Energy for Immiscible Two-Fluid Flow in Porous Media by Integral Geometry and Thermodynamics. *Water Resour. Res.* **2018**, *54*, 9045–9059. [CrossRef]
11. Armstrong, R.T.; McClure, J.E.; Robins, V.; Liu, Z.; Arns, C.H.; Schlüter, S.; Berg, S. Porous media characterization using minkowski functionals: Theories, applications and future directions. *Transp. Porous Med.* **2019**, *130*, 305–335. [CrossRef]
12. Slotte, P.A.; Berg, C.F.; Khanamiri, H.H. Predicting Resistivity and Permeability of Porous Media Using Minkowski Functionals. *Transp. Porous Med.* **2020**, *131*, 705–722. [CrossRef]
13. Kjelstrup, S.; Bedeaux, D.; Hansen, A.; Hafskjold, B.; Galteland, O. Non-isothermal transport of multi-phase fluids in porous media. the entropy production. *Front. Phys.* **2018**, *6*, 126. [CrossRef]
14. Kjelstrup, S.; Bedeaux, D.; Hansen, A.; Hafskjold, B.; Galteland, O. Non-isothermal transport of multi-phase fluids in porous media. Constitutive equations. *Front. Phys.* **2019**, *6*, 150. [CrossRef]
15. Balbuena, P.B.; Berry, D.; Gubbins, K.E. Solvation pressures for simple fluids in micropores. *J. Phys. Chem.* **1993**, *97*, 937–943. [CrossRef]
16. Gubbins, K.E.; Long, Y.; Śliwinska-Bartkowiak, M. Thermodynamics of confined nano-phases. *J. Chem. Thermodyn.* **2014**, *74*, 169–183. [CrossRef]
17. Bedeaux, D.; Kjelstrup, S. Hill's nano-thermodynamics is equivalent with Gibbs' thermodynamics for surfaces of constant curvatures. *Chem. Phys. Lett.* **2018**, *707*, 40–43. [CrossRef]
18. Gjennestad, M.A.; Wilhelmsen, Ø. Thermodynamic stability of volatile droplets and thin films governed by the disjoining pressure in open and closed containers. *Langmuir* **2020**, *36*, 27. [CrossRef]
19. Gjennestad, M.A.; Wilhelmsen, Ø. Thermodynamic stability of droplets, bubbles and thick films in open and closed pores. *Fluid Phase Equilibr.* **2020**, *505*, 112351. [CrossRef]
20. Strøm, B.A.; Simon, J.M.; Schnell, S.K.; Kjelstrup, S.; He, J.; Bedeaux, D. Size and shape effects on the thermodynamic properties of nanoscale volumes of water. *Phys. Chem. Chem. Phys.* **2017**, *19*, 9016–9027. [CrossRef]

21. Galteland, O.; Bedeaux, D.; Hafskjold, B.; Kjelstrup, S. Pressures inside a nano-porous medium. The case of a single phase fluid. *Front. Phys.* **2019**, *7*, 60. [CrossRef]

22. Erdős, M.; Galteland, O.; Bedeaux, D.; Kjelstrup, S.; Moultos, O.A.; Vlugt, T.J. Gibbs Ensemble Monte Carlo Simulation of Fluids in Confinement: Relation between the Differential and Integral Pressures. *Nanomaterials* **2020**, *10*, 293. [CrossRef]

23. Rauter, M.T.; Galteland, O.; Erdős, M.; Moultos, O.A.; Vlugt, T.J.; Schnell, S.K.; Bedeaux, D.; Kjelstrup, S. Two-Phase Equilibrium Conditions in Nanopores. *Nanomaterials* **2020**, *10*, 608. [CrossRef] [PubMed]

24. Strøm, B.A.; He, J.; Bedeaux, D.; Kjelstrup, S. When Thermodynamic Properties of Adsorbed Films Depend on Size: Fundamental Theory and Case Study. *Nanomaterials* **2020**, *10*, 1691. [CrossRef] [PubMed]

25. Bedeaux, D.; Kjelstrup, S.; Schnell, S.K. *Nanothermodynamics. General Theory*; NTNU: Trondheim, Norway, 2020.

26. Hill, T.L. *Thermodynamics of Small Systems - Two Volumes Bound as One*; Dover: New York, NY, USA, 1964.

27. Hill, T.L.; Chamberlin, R.V. Extension of the thermodynamics of small systems to open metastable states: An example. *Proc. Natl. Acad. Sci. USA* **1998**, *95*, 12779–12782. [CrossRef] [PubMed]

28. Hill, T.L.; Chamberlin, R.V. Fluctuations in energy in completely open small systems. *Nano Lett.* **2002**, *2*, 609–613. [CrossRef]

29. Hansen, J.P.; McDonald, I.R. *Theory of Simple Liquids*; Elsevier: Amsterdam, The Netherlands, 1990.

30. Radke, C. Film and membrane-model thermodynamics of free thin liquid films. *J. Colloid Interf. Sci.* **2015**, *449*, 462–479. [CrossRef]

31. Long, Y.; Palmer, J.C.; Coasne, B.; Śliwinska-Bartkowiak, M.; Gubbins, K.E. Pressure enhancement in carbon nanopores: A major confinement effect. *Phys. Chem. Chem. Phys.* **2011**, *13*, 17163–17170. [CrossRef]

32. van Dijk, D. Comment on "Pressure enhancement in carbon nanopores: A major confinement effect" by Y. Long, J. C. Palmer, B. Coasne, M. Sliwinska-Bartkowiak and K. E. Gubbins, Phys. Chem. Chem. Phys., 2011, 13, 17163. *Phys. Chem. Chem. Phys.* **2020**, *22*, 9824–9825. [CrossRef]

33. Long, Y.; Palmer, J.C.; Coasne, B.; Shi, K.; Śliwińska-Bartkowiak, M.; Gubbins, K.E. Reply to the 'Comment on "Pressure enhancement in carbon nanopores: A major confinement effect"'by D. van Dijk, Phys. Chem. Chem. Phys., 2020, 22. *Phys. Chem. Chem. Phys.* **2020**, *22*, 9826–9830. [CrossRef]

34. Schofield, P.; Henderson, J.R. Statistical mechanics of inhomogeneous fluids. *Proc. R. Soc. Lon. Ser. A* **1982**, *379*, 231–246.

35. Harasima, A. Molecular theory of surface tension. *Adv. Chem. Phys.* **1958**, *1*, 203–237.

36. Irving, J.; Kirkwood, J.G. The statistical mechanical theory of transport processes. IV. The equations of hydrodynamics. *J. Chem. Phys.* **1950**, *18*, 817–829. [CrossRef]

37. Hafskjold, B.; Ikeshoji, T. Microscopic pressure tensor for hard-sphere fluids. *Phys. Rev. E* **2002**, *66*, 011203. [CrossRef] [PubMed]

38. Shi, K.; Shen, Y.; Santiso, E.E.; Gubbins, K.E. Microscopic pressure tensor in cylindrical geometry: Pressure of water in a carbon nanotube. *J. Chem. Theory Comput.* **2020**, *16*, 5548–5561. [CrossRef] [PubMed]

39. Ikeshoji, T.; Hafskjold, B.; Furuholt, H. Molecular-level calculation scheme for pressure in inhomogeneous systems of flat and spherical layers. *Mol. Simulat.* **2003**, *29*, 101–109. [CrossRef]

40. Evans, R.; Marini Bettolo Marconi, U. Phase equilibria and solvation forces for fluids confined between parallel walls. *J. Chem. Phys.* **1987**, *86*, 7138–7148. [CrossRef]

41. Frenkel, D.; Smit, B. *Understanding Molecular Simulation: From Algorithms to Applications*; Elsevier: Amsterdam, The Netherlands, 2001; Volume 1.

42. Shinoda, W.; Shiga, M.; Mikami, M. Rapid estimation of elastic constants by molecular dynamics simulation under constant stress. *Phys. Rev. B* **2004**, *69*, 134103. [CrossRef]

43. Plimpton, S. Fast parallel algorithms for short-range molecular dynamics. *J. Comput. Phys.* **1995**, *117*, 1–19. [CrossRef]

44. Hafskjold, B.; Travis, K.P.; Hass, A.B.; Hammer, M.; Aasen, A.; Wilhelmsen, Ø. Thermodynamic properties of the 3D Lennard-Jones/spline model. *Mol. Phys.* **2019**, *117*, 3754–3769. [CrossRef]

45. Stukowski, A. Visualization and analysis of atomistic simulation data with OVITO–the Open Visualization Tool. *Model. Simul. Mater. Sci.* **2009**, *18*, 015012. [CrossRef]

 *nanomaterials*

Article

A Legendre–Fenchel Transform for Molecular Stretching Energies

Eivind Bering [1,*], **Dick Bedeaux** [2,*], **Signe Kjelstrup** [2], **Astrid S. de Wijn** [3], **Ivan Latella** [4] **and J. Miguel Rubi** [4]

[1] PoreLab, Department of Physics, Norwegian University of Science and Technology,
     NO–7491 Trondheim, Norway
[2] PoreLab, Department of Chemistry, Norwegian University of Science and Technology,
     NO–7491 Trondheim, Norway; signe.kjelstrup@ntnu.no
[3] Department of Mechanical and Industrial Engineering, Norwegian University of Science and Technology,
     NO–7491 Trondheim, Norway; astrid.dewijn@ntnu.no
[4] Department of Condensed Matter Physics, Universitat de Barcelona, Av.Diagonal 647,
     08028 Barcelona, Spain; ilatella@ub.edu (I.L.); mrubi@ub.edu (J.M.R.)
*   Correspondence: eivind.bering@ntnu.no (E.B.); dick.bedeaux@ntnu.no (D.B.)

Received: 10 November 2020; Accepted: 25 November 2020; Published: 27 November 2020

Abstract: Single-molecular polymers can be used to analyze to what extent thermodynamics applies when the size of the system is drastically reduced. We have recently verified using molecular-dynamics simulations that isometric and isotensional stretching of a small polymer result in Helmholtz and Gibbs stretching energies, which are not related to a Legendre transform, as they are for sufficiently long polymers. This disparity has also been observed experimentally. Using molecular dynamics simulations of polyethylene-oxide, we document for the first time that the Helmholtz and Gibbs stretching energies can be related by a Legendre–Fenchel transform. This opens up a possibility to apply this transform to other systems which are small in Hill's sense.

Keywords: nanothermodynamics; polymers; molecular simulation; single-molecule stretching

---

1. Introduction

As we reduce system dimensions from the micro- to the nano-scale, surface properties become increasingly important, and the normal thermodynamic equations (thermodynamic limit properties) cease to apply. Hill [1] proposed a way to restore the structure of ordinary Gibbs' thermodynamics to deal with small systems. His idea was to introduce an ensemble of small systems, for which ordinary thermodynamics again can be applied. For an in-depth discussion, see also [2]. In Hill's description, Legendre transforms and Maxwell relations exist, but only at the level of the ensemble of small systems. A single small system, however, does not obey the normal Legendre transforms. A characteristic of small systems is that extensive properties cease to be extensive due to finite size effects, and the thermodynamic potentials depend on the type of environmental control variables, or the ensemble to which they belong. In other words, in general, statistical ensembles are not equivalent for small systems. This striking property is typically observed also in systems with size comparable with the range of the interactions [3–5]. Ensemble inequivalence in long-range interacting systems is related to the occurrence of curvature anomalies in thermodynamic potentials, which in this case arise because the interaction energy is not additive. It has been shown that Hill's approach for small systems can be implemented for long-range interacting systems as well, and that it naturally takes into account the non-additivity induced by the interactions [6,7]. Such a parallelism between small systems and systems with long-range interactions [8,9] indicates that the methods used to describe long-range interacting systems also may find a wider application in the characterization of small systems, and vice versa.

For small systems, the relative size of the fluctuations will be of more significance than for a typical large system. For sufficiently small polymers with a non-linear force-response, one would expect the difference in fluctuations to give rise to size-dependent ensemble deviations. The energy involved in stretching then depends on whether one controls the stretching length or the stretching force. The average force for isometric stretching of a small molecule differs from that of isotensional stretching. In the long polymer limit, they are the same, however, and this has been verified experimentally, computationally, and theoretically. A detailed discussion of this is given by Süzen et al. [10].

We have also studied this problem [11], and verified that the forces were not the same, as predicted from theory. This resulted in a Helmholtz energy for isometric stretching and a Gibbs energy for isotensional stretching for small molecules that were not related by a Legendre transform, which is also known from experiments by Keller et al. [12]. In addition, ensemble inequivalence has been recently highlighted in pulling experiments by Monge et al. [13].

A question therefore arises: is it then at all possible to transform the small system description from one set of variables to another set, like we normally do when we use Legendre transforms? To be more specific: is it possible to transform the Helmholtz energy of a molecule (which describes isometric stretching) into its Gibbs energy (which applies for isotensional stretching)? The aim of this short communication is to show that this is indeed possible.

We shall use our earlier simulation results [11] and verify that the Helmholtz and Gibbs energies for the stretching of a short polymer can be related to each other using the Legendre–Fenchel transform [14], a generalization of the usual Legendre transform, suitable for free energies that exhibit curvature anomalies. This transform has already proven useful in long-range interacting systems displaying ensemble inequivalence [5,15,16], and here it is applied for the first time to a common stretching phenomenon. The Legendre–Fenchel transform reduces to the usual Legendre transform when the Helmholtz energy is differentiable and convex; in the present case, this happens for large polymers. As we precisely show with our numerical simulations, the Helmoltz energy of the considered small polymers in fact present curvature anomalies under certain conditions, making it impossible to use the conventional Legendre transform.

2. Method

This section is split into two parts. The first part introduces the model and the computational details, and the second part presents the theoretical method.

2.1. Simulation Details

We use the same model as some of us have used previously [11,17] to investigate molecular stretching of poly-ethylene oxide (PEO) on the form $CH_3-[O-CH_2-CH_2]_n-O-CH_3$ in molecular dynamics simulations. It is a united-atom model with each bead representing either a methyl group, a methylene group or an oxygen atom. This model is based on a common model documented in the literature [18–20], and has all the standard contributions to the potential energy from bond stretching, bending, and torsion, and includes also the breaking of bonds. It therefore lends itself well to a testing of the stretching energies. In this particular force-field, the standard harmonic bond stretching potential is replaced by a Morse potential

$$U_{\text{bond}}(\{\mathbf{R_i}, i = 1, N\}) = D_{ij} \left[1 - e^{-\alpha_{ij}(r_{ij}-\bar{r}_{ij})}\right]^2,\tag{1}$$

where the parameters for the dissociation energies D_{ij} are obtained from density functional computations from the literature [21]. The stiffness of the bond is determined by $\alpha_{ij} = \sqrt{K_{ij}^s/2D_{ij}}$. Furthermore, the potentials for the bending and torsion of bonds read

$$U_{\text{bend}}(\{\mathbf{R_i}\}) = \frac{1}{2} \sum_{\{ijk\}} K^b_{ijk}[\theta_{ijk} - \bar{\theta}_{ijk}]^2 \tag{2}$$

and

$$U_{\text{tors}}(\{\mathbf{R_i}\}) = \sum_{\{ijkl\}} \sum_{\{c\}} K^{t,c}_{ijkl} \cos^{c-1}(\phi_{ijkl}), \tag{3}$$

where i, j, k and l are atoms joined by consecutive covalent bonds. K^s_{ij}, K^b_{ijk}, K^t_{ijkl} are force constants for stretching (s), bending (b) and torsion (t). $\bar{r}_{ij}$ and $\bar{\theta}_{ijk}$ are equilibrium values for bond stretching and bending, respectively. All force-field parameters were tabulated previously [11]. Non-bonded interactions were not taken into account in the current work, which means that our model polymer is surrounded by an implicit theta solvent. The force field is compatible with the LAMMPS [22] simulation package, that has been used for all of our computations.

The temperature was set to 300 K during sampling, and was controlled by a Langevin thermostat with a relaxation time of 1 ps and a time step of 1 fs. The initial configurations were exposed to a simulated annealing protocol prior to sampling, in an attempt to capture a representative portion of the phase space [17,23]. The presented data are averaged over 5 ns for 200 samples.

2.2. Energy Transforms

For the theoretical analysis, consider now an arbitrary polymer with N beads. The energy of the polymer is given by

$$H(\mathbf{r}_1, ..., \mathbf{r}_N; \mathbf{p}_1, ..., \mathbf{p}_N) = \sum_{j=1}^{N} \frac{p_j^2}{2m_j} + V(\mathbf{r}_1, ..., \mathbf{r}_N), \tag{4}$$

where $p_j \equiv |\mathbf{p}_j|$, m_j is the mass of bead j, and $V(\mathbf{r}_1, ..., \mathbf{r}_N)$ is the potential interaction. In our previous work [11,17], we gave an explicit expression for the interaction potential with contributions from bond stretching, bending, and torsion. The polymer is controlled either in the isometric ensemble by fixing the end-to-end distance $x \equiv |\mathbf{r}_N - \mathbf{r}_1|$, or in the isotensional ensemble by applying a stretching force $f \equiv |\mathbf{f}_N - \mathbf{f}_1|$. The canonical partition function in the isometric ensemble is

$$Z(T, N, x) = \frac{1}{\hbar^{3(N-1)} N!} \int' d\mathbf{r}_1 ... d\mathbf{r}_N \int' d\mathbf{p}_1 ... d\mathbf{p}_N \exp(-\beta H), \tag{5}$$

where the end-to-end distance x is controlled, by keeping $\mathbf{r}_N - \mathbf{r}_1$ constant in the integral over the spacial coordinates. The prime for the spacial integrals indicates this. The prime for the momenta indicates that we keep the center of mass fixed. Furthermore $\hbar$ is Planck's constant and $\beta \equiv 1/(k_B T)$, where k_B is Boltzmann's constant. Because of the symmetry of the system the partition function Z depends only on x and not on the direction of $\mathbf{r}_N - \mathbf{r}_1$. The partition function for the isotensional ensemble is

$$\Delta(T, N, f) = \beta f \int_0^{x_{\max}} dx\, Z(T, x) \exp(\beta f x), \tag{6}$$

where now the stretching force is constant, and $x_{\max}$ denotes the length of the unfolded polymer. The Helmholtz energy is given by

$$F(T, N, x) = -k_B T \ln Z(T, N, x), \tag{7}$$

and the Gibbs energy by

$$G(T, N, f) = -k_B T \ln \Delta(T, N, f), \tag{8}$$

in which x and f are the relevant conjugated variables as usully considered in thermodynamics and statistical mechanics of polymer systems [24]. It follows from Equation (6) that

$$\exp[-\beta G(T, N, f)] = \beta f \int_0^{x_{\max}} dx \exp\{-\beta[F(T, N, x) - fx]\}. \tag{9}$$

This makes it possible to calculate the Gibbs energy in the isotensional ensemble from the Helmholtz energy in the isometric ensemble. As the above derivation shows, this transformation is also correct for small polymers.

For sufficiently long polymers, the usual Legendre transform

$$G\left(T, N, f(x)\right) = F(T, N, x) - f(x)x \tag{10}$$

is valid. However, we verified in our first paper [11], using molecular dynamics simulations, that for small polymers, the usual Legendre transform is not valid.

Differences in the Helmholtz energy are calculated using

$$F(T, N, x_1) - F(T, N, x_0) = \int_{x_0}^{x_1} \bar{f}(x)dx \, . \tag{11}$$

Gibbs energy differences are calculated using

$$G(T, N, f_1) - G(T, N, f_0) = -\int_{f_0}^{f_1} \bar{x}(f)df \, . \tag{12}$$

By Equation (9), one may also find the Gibbs energy from $F(T, N, x_1)$ in Equation (11). With $x_0 = 0$ and $f_0 = 0$, Equation (9) gives

$$\exp\left\{-\beta G\left(T, N, f(x_1)\right)\right\} = \beta f(x_1) \int_0^{x_{\max}} dx \exp\left\{-\beta\left[F(T, N, x) - f(x_1)x\right]\right\} \, , \tag{13}$$

where $f(x_1)$ is obtained by means of interpolation of the isotensional force-elongation curve. From a saddle point approximation to compute the integral in Equation (13), one obtains

$$-G_{\mathrm{LF}}\left(T, N, f(x_1)\right) = \max_x \left[f(x_1)x - F(T, N, x)\right] \, . \tag{14}$$

The function $F^*(T, N, f) = -G_{\mathrm{LF}}(T, N, f)$ is known as the Legendre–Fenchel transform [5,15,16] of $F(T, N, x)$ with respect to x at constant T and N.

The Legendre–Fenchel transform is a generalization of the Legendre transform, well known in statistical physics [5,16], and reduces to the latter when the transformed function is differentiable and convex. An important property of the Legendre–Fenchel transform is that it always yields convex functions; thus $-G_{\mathrm{LF}}(f)$ is convex in f at constant T and N. Furthermore, if $F^*(f) = -G_{\mathrm{LF}}(f)$ is transformed again, one has

$$F^{**}(x) = \max_f [fx + G_{\mathrm{LF}}(f)]. \tag{15}$$

Because $-G_{\mathrm{LF}}(f)$ is a convex function, at points f for which $-G_{\mathrm{LF}}(f)$ is differentiable the above transform (15) reduces to the usual Legendre transform, leading to

$$F^{**}(x) = f(x)x + G_{\mathrm{LF}}(f(x)), \tag{16}$$

where $f(x)$ is the unique solution to $dG_{\mathrm{LF}}(f)/df = -x$. Since $F^{**}(x)$ is simply the Legendre transform of $-G_{\mathrm{LF}}(f)$, the former is the isotensional Helmholtz free energy. Moreover, due to the properties of the Legendre–Fencel transform, F^{**} is the convex envelope of the isometric free energy F, namely, the largest convex function such that $F^{**} \leq F$. Thus, the isometric and isotensional ensembles are not equivalent if F does not coincide with its convex envelope F^{**}. In mathematical terms, ensemble inequivalence may arise because the Legendre–Fenchel transform is not necessarily self-dual (or involute), that is, $F^{**} \neq F$ when F is non-convex. In contrast, the convex envelope F^{**} of F has the same Legendre–Fenchel transform as F, meaning that $(F^{**})^* = F^*$ [16].

We have outlined a method for obtaining the free energy G_{LF} in the isotensional ensemble from the free energy F in the isometric ensemble. This method applies, in particular, when F is non-convex. By computing the derivative of $G_{LF} = G_{LF}(T, N, f(x_1))$ with respect to f, one obtains the force elongation relation

$$\frac{d}{df} G_{LF} = -x(f) \tag{17}$$

in the isotensional ensemble. The purpose of this paper is now to test these formulas.

3. Simulation Results

In the molecular dynamics simulations, one obtains the average force $\bar{f}(x) = \langle f(t) \rangle_x$ between the end points in the isometric ensemble. In the isotensional ensemble, one obtains the average distance between the endpoints $\bar{x}(f) = \langle x(t) \rangle_f$. The force-elongation curves from the isometric and isotensional ensembles are shown as a function of the length per bond $x_b = x/(N-1)$ for systems of size $N = 12, 24$ and 51 in Figure 1a–c.

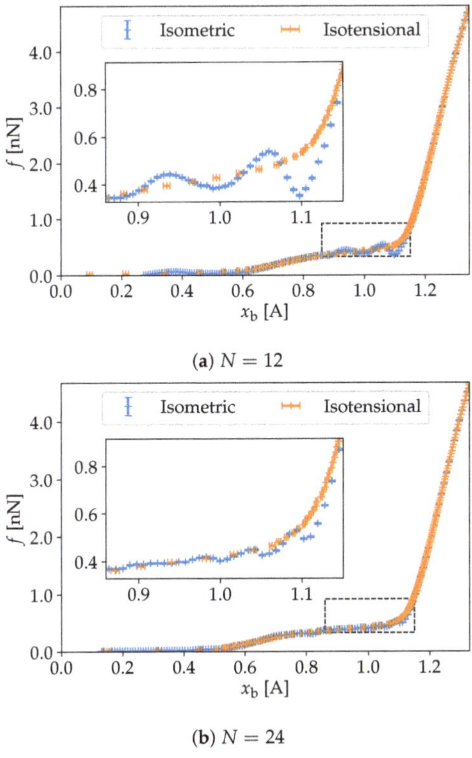

(**a**) $N = 12$

(**b**) $N = 24$

Figure 1. *Cont.*

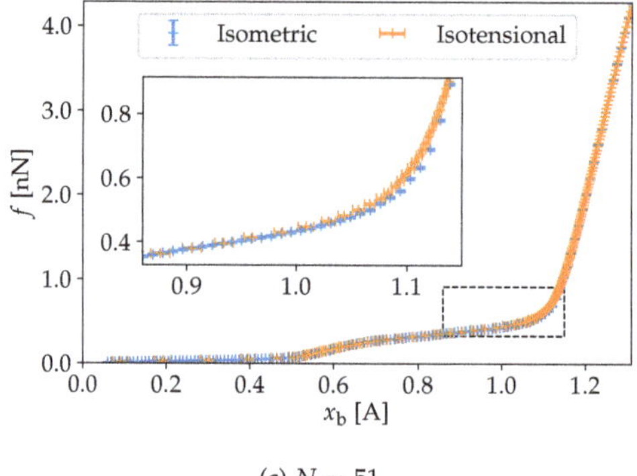

(c) $N = 51$

Figure 1. Force as a function of length per bond from isometric and isotensional simulations for chains of poly-ethylene oxide (PEO) composed of $N = 12, 24$ and 51 united atoms. The ensemble inequivalence is most pronounced for the smallest systems.

It is clear from these figures that the isometric and the isotensional force are different, a fact that is more pronounced for the smaller polymers. In the isometric ensemble, the slope of the curve $\bar{f}(x)$ is not restricted to be a positive quantity, since the Helmoltz free energy $F(T, N, x)$ is not necessarily a convex function with respect to x at fixed T and N.

In other words, the response function $\kappa^{(x)}$ defined through

$$\frac{1}{\kappa^{(x)}} = \left(\frac{\partial \bar{f}}{\partial x}\right)_{T,N} = \left(\frac{\partial^2 F}{\partial x^2}\right)_{T,N} \tag{18}$$

can be negative in the isometric ensemble [25], meaning that the associated system configurations minimize the free energy when the average force between the ends of the polymer decreases for increasing elongation. Under these conditions, interactions between monomers tend to separate them from each other, decreasing internal forces required to keep the polymer in equilibrium. We highlight that negative values of $\kappa^{(x)}$ in this ensemble may be realized because x is always kept fixed at a definite value. Furthermore, in the isotensional ensemble, the end-to-end distance fluctuates at constant applied force. In that case, the slope of $\bar{x}(f)$ cannot be negative, namely,

$$\kappa^{(f)} = \left(\frac{\partial \bar{x}}{\partial f}\right)_{T,N} = -\left(\frac{\partial^2 G}{\partial f^2}\right)_{T,N} \geq 0, \tag{19}$$

because internal forces under these conditions do not equilibrate with the external force applied on the polymer.

The points of negative slope in the isometric ensemble can be explained by the torsional unfolding of the molecule. These mechanically unstable modes are not accessible in the isotensional ensemble. As a consequence, we see that the ensemble deviation is most pronounced around $x_b = 1.1$, which marks the end of the region for torsional unfolding. This was previously discussed in great detail [11]. Prior to this region, around $x_b < 0.5$, the molecule is twisted helically, and the relation between force and elongation is predominantly linear due to entropic effects. In the last regime, with $x_b > 1.1$, the molecule is planar, and the force-elongation curve is dominated by the stretching of the individual monomers.

Differences in the Helmoltz energy are found from the isometric ensembles by Equation (11), and the Gibbs energy differences are found from the isotensional ensemble by Equation (12). The Legendre–Fenchel transform of the Helmoltz energy is then found by Equation (14). We present these curves for systems of size $N = 12, 24$ and 51 in Figure 2a–c. The Gibbs energy is shown with an orange line, and is compared to the Legendre transform of the Helmholtz energy in blue. It is clear that the Legendre–Fenchel transform of the Helmholtz energy, shown with a black dotted line, gives an approximation to the Gibbs energy that is far superior that of the Legendre transform. We would also like to stress that the Legendre–Fenchel transform is exact in the limit $N \to \infty$, since the saddle-point approximation is exact in this limit.

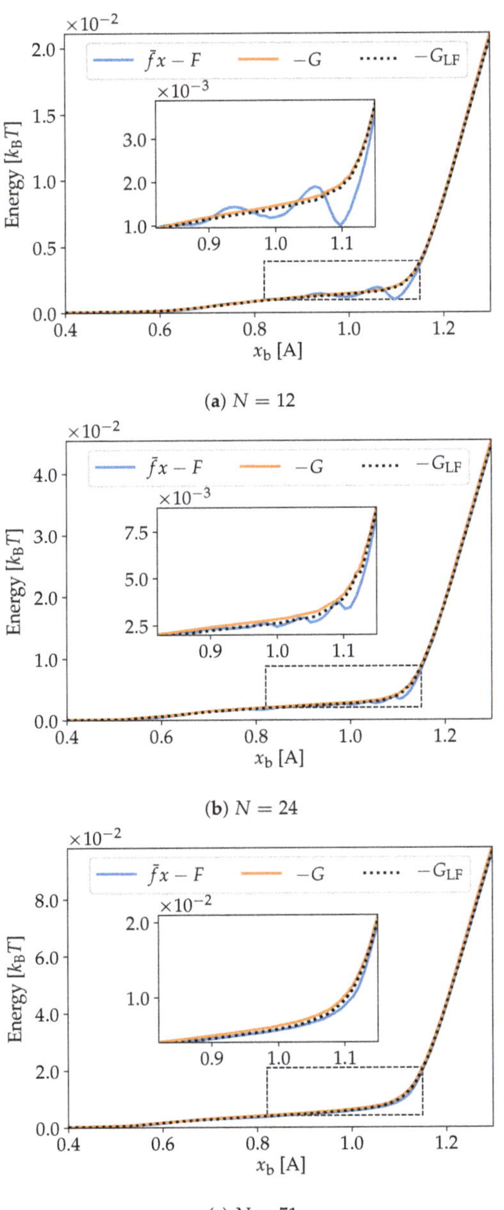

(**a**) $N = 12$

(**b**) $N = 24$

(**c**) $N = 51$

Figure 2. Energy as a function of length per bond for chains of PEO composed of $N = 12$, 24 and 51 united atoms. While the Legendre transform of the Helmholtz energy F is different from minus the Gibbs energy G, we see that the Legendre–Fenchel transform G_{LF} is an excellent approximation in all three cases.

We can see from Figure 2a–c that even for finite N, the free energy $G_{LF}\,(T, N, f(x_1))$ is in excellent approximation equal to $G\,(T, N, f(x_1))$. This shows that the exact transformation, which follows from the relation between the partition function, given in Equation (13), as well as the approximate Legendre–Fenchel transform, Equation (14), can be used to obtain the Gibbs energy from the Helmholtz energy for the stretching of small polymers. The curve for $\bar{f}(x_1)x_1 - F(T, N, x_1)$ is the result of the isometric simulations and differs from the Gibbs energy curves. This shows clearly that the Legendre transform, given in Equation (10), is not valid for small polymers.

In Figure 3a,b we present the energies from Figure 2a,b as a function of force rather than elongation for systems of size $N = 12$ and 24. As the curves for the Helmholtz energy for these systems are not convex, the corresponding Legendre transformed curves as a function of force is not one-to-one. This is emphasized in the inserts.

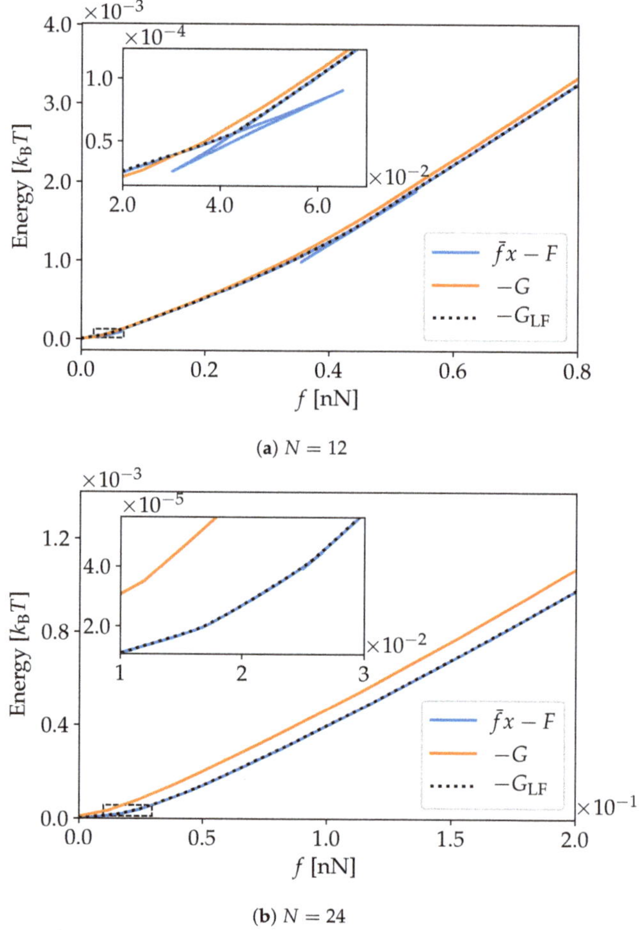

(**a**) $N = 12$

(**b**) $N = 24$

Figure 3. Energy as a function of force for chains of PEO of composed of $N = 12$ and 24 united atoms. The smallest system displays multiple singularities, one of which is emphasized in the insert. Although less pronounced, singularities can be seen also in the system with $N = 24$.

The force elongation relation for the Legendre–Fenchel transformed energy can be obtained by the derivative of G_{LF} with respect to f, cf. Equation (17). This monotonically increasing curve is shown with a black dotted line in Figure 4a–c, with the original force-elongation curves for comparison. It is clear that the Legendre–Fenchel transform is non-involutive for $N = 12$, and that it is involutive for $N = 51$, where it reduces to the Legendre transform [14].

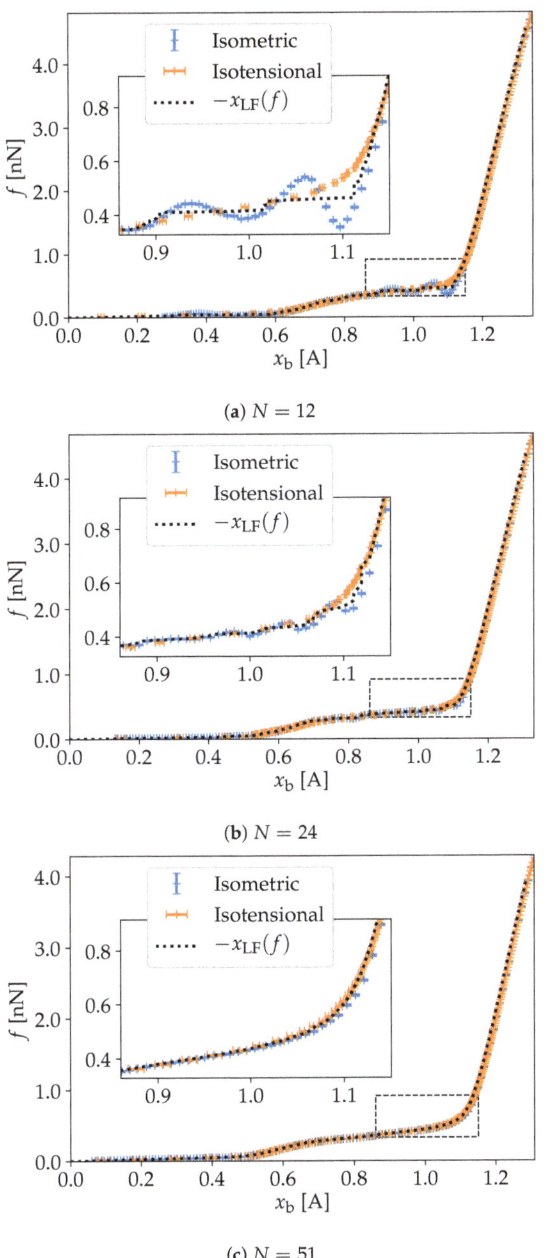

(**a**) $N = 12$

(**b**) $N = 24$

(**c**) $N = 51$

Figure 4. The force-elongation curve x_{LF} computed from the Legendre–Fenchel transform cf. Equation (17) is compared to the force-elongation curves from Figure 1. We recognize the singular points in G_{LF} as jumps in $x_{\mathrm{LF}}(f)$, particularly visible in the smallest system with $N = 12$.

4. Discussion and Conclusions

We have analyzed the stretching of small polymers in which the thermodynamic limit cannot be invoked. We have shown that small size contributions to the isometric Helmholtz free energy induce curvature anomalies in this thermodynamic potential, which disappear as the number of beads in the polymer is increased. We described a method employing the Legendre–Fenchel transform to manage these curvature anomalies and obtain the isotensional Gibbs free energy from simulations in the isometric ensemble, in such a way that the states characterized by this free energy are unique.

The Legendre–Fenchel transform in Equation (14) reduces to the usual Legendre transform (Equation (10)) when the free energy $F(x)$ is differentiable and convex in x at constant T and N. Legendre–Fenchel transforms rather than Legendre transforms must be used in particular because $F(x)$ is non-convex [5,16]. As noted previously, the Legendre–Fenchel transform always yields convex functions and therefore, $-G_{\mathrm{LF}}(f)$ is convex in f at constant T and N. The fact that $-G_{\mathrm{LF}}(f)$ is convex ensures that the slope of the curve $\bar{x}(f)$ is non-negative, as required in equilibrium states under fluctuations of the end-to-end distance. Remarkably, this is the case when the free energy $F(x)$ presents a non-convex anomaly in the isometric ensemble. This implies a negative slope in the curve $\bar{f}(x)$. The Legendre–Fenchel transform maps the states associated with the anomaly into a point f at which $-G_{\mathrm{LF}}(f)$ is non-differentiable. This behavior is exemplified in Figure 3a,b for $N = 12$ and $N = 24$, respectively; in particular in the inserts. Such singularities are not observed for $N = 51$, as $F(x)$ in this case is convex.

We have seen above that the Legendre–Fenchel transform enables us to transform the stretching energy from the isometric to the isotensional ensemble also for small polymers. This removes the limitations set by the Legendre transforms, applicable only in the thermodynamic limit, and opens up a possibility for wider applications. The scheme documented here for molecular stretching energies reduces to the usual Legendre transforms in the thermodynamic limit.

Author Contributions: E.B. contributed software, validation and formal analysis, with A.S.d.W. as supervisor. Conceptualization was by D.B., I.L. and J.M.R. D.B., S.K. and E.B. wrote the original draft. All authors contributed to the writing—review. All authors have read and agreed to the published version of the manuscript.

Funding: This research was funded by the Research Council of Norway grant no. 262644 (PoreLab Center of Excellence) and 250158 (yield kinetics).

Acknowledgments: The IDUN cluster at NTNU HPC Group is acknowledged for computational resources. Some computations were also performed on resources provided by UNINETT Sigma2—the National Infrastructure for High Performance Computing and Data Storage in Norway.

Conflicts of Interest: The authors declare no conflict of interest. The funders had no role in the design of the study; in the collection, analyses, or interpretation of data; in the writing of the manuscript, or in the decision to publish the results.

References

1. Hill, T.L. *Thermodynamics of Small Systems*; W.A. Benjamin Inc.: New York, NY, USA, 1963; p. 210.
2. Bedeaux, D.; Kjelstrup, S.; Schnell, S.K. *Nanothermodynamics—General Theory*, 1st ed.; PoreLab: Trondheim, Norway, 2020.
3. Campa, A.; Dauxois, T.; Ruffo, S. Statistical mechanics and dynamics of solvable models with long-range interactions. *Phys. Rep.* **2009**, *480*, 57–159. [CrossRef]
4. Bouchet, F.; Gupta, S.; Mukamel, D. Thermodynamics and dynamics of systems with long-range interactions. *Phys. A Stat. Mech. Its Appl.* **2010**, *389*, 4389–4405. [CrossRef]
5. Campa, A.; Dauxois, T.; Fanelli, D.; Ruffo, S. *Physics of Long-Range Interacting Systems*; Oxford University Press: Oxford, UK, 2014.
6. Latella, I.; Pérez-Madrid, A. Local thermodynamics and the generalized Gibbs-Duhem equation in systems with long-range interactions. *Phys. Rev. E* **2013**, *88*, 42135. [CrossRef]
7. Latella, I.; Pérez-Madrid, A.; Campa, A.; Casetti, L.; Ruffo, S. Thermodynamics of Nonadditive Systems. *Phys. Rev. Lett.* **2015**, *114*, 230601. [CrossRef]
8. Gross, D.H.E. *Microcanonical Thermodynamics*; World Scientific: Singapore, 2001.

9. Chomaz, P.; Gulminelli, F. Phase Transitions in Finite Systems. In *Dynamics and Thermodynamics of Systems with Long Range Interactions*; Dauxois, T., Ruffo, S., Arimondo, E., Wilkens, M., Eds.; Springer: Berlin/Heidelberg, Germany, 2002; Volume 602, pp. 68–129.

10. Süzen, M.; Sega, M.; Holm, C. Ensemble inequivalence in single-molecule experiments. *Phys. Rev. E* **2009**, *79*, 051118. [CrossRef]

11. Bering, E.; Kjelstrup, S.; Bedeaux, D.; Miguel Rubi, J.; de Wijn, A.S. Entropy Production beyond the Thermodynamic Limit from Single-Molecule Stretching Simulations. *J. Phys. Chem. B* **2020**, *124*, 8909–8917. [CrossRef] [PubMed]

12. Keller, D.; Swigon, D.; Bustamante, C. Relating Single-Molecule Measurements to Thermodynamics. *Biophys. J.* **2003**, *84*, 733–738. [CrossRef]

13. Monge, A.M.; Manosas, M.; Ritort, F. Experimental test of ensemble inequivalence and the fluctuation theorem in the force ensemble in DNA pulling experiments. *Phys. Rev. E* **2018**, *98*, 032146. [CrossRef]

14. Rockafellar, T. *Convex Analysis*; Princeton University Press: Princeton, NJ, USA, 1972.

15. Ellis, R.; Haven, K.; Turkington, B. Large Deviation Principles and Complete Equivalence and Nonequivalence Results for Pure and Mixed Ensembles. *J. Stat. Phys.* **2000**, *101*, 999–1064. [CrossRef]

16. Touchette, H. The large deviation approach to statistical mechanics. *Phys. Rep.* **2009**, *478*, 1–69. [CrossRef]

17. Bering, E.; de Wijn, A.S. Stretching and breaking of PEO nanofibres. A classical force field and ab initio simulation study. *Soft Matter* **2020**, *16*, 2736–2752. [CrossRef] [PubMed]

18. Chen, C.; Depa, P.; Sakai, V.G.; Maranas, J.K.; Lynn, J.W.; Peral, I.; Copley, J.R.D. A comparison of united atom, explicit atom, and coarse-grained simulation models for poly(ethylene oxide). *J. Chem. Phys.* **2006**, *124*, 234901(11). [CrossRef] [PubMed]

19. van Zon, A.; Mos, B.; Verkerk, P.; de Leeuw, S. On the dynamics of PEO-NaI polymer electrolytes. *Electrochim. Acta* **2001**, *46*, 1717–1721. [CrossRef]

20. Neyertz, S.; Brown, D.; Thomas, J.O. Molecular dynamics simulation of crystalline poly(ethylene oxide). *J. Chem. Phys.* **1994**, *101*, 10064. [CrossRef]

21. Beyer, M.K. The mechanical strength of a covalent bond calculated by density functional theory. *J. Chem. Phys.* **2000**, *112*, 7307–7312. [CrossRef]

22. Plimpton, S. Fast parallel algorithms for short-range molecular dynamics. *J. Comput. Phys.* **1995**, *117*, 1–19. [CrossRef]

23. Steinbach, P.J.; Brooks, B.R. Protein simulation below the glass-transition temperature. Dependence on cooling protocol. *Chem. Phys. Lett.* **1994**, *226*, 447–452. [CrossRef]

24. Hill, T.L. *An Introduction to Statistical Thermodynamics*; Dover Publications Inc.: New York, NY, USA, 1986.

25. Campa, A.; Casetti, L.; Latella, I.; Pérez-Madrid, A.; Ruffo, S. Concavity, Response Functions and Replica Energy. *Entropy* **2018**, *20*, 907. [CrossRef]

Publisher's Note: MDPI stays neutral with regard to jurisdictional claims in published maps and institutional affiliations.

 *nanomaterials*

Article

When Thermodynamic Properties of Adsorbed Films Depend on Size: Fundamental Theory and Case Study

Bjørn A. Strøm [1,*], **Jianying He** [1], **Dick Bedeaux** [2] **and Signe Kjelstrup** [2]

[1] Department of Structural Engineering, Faculty of Engineering Science and Technology, Norwegian University of Science and Technology, NO-7491 Trondheim, Norway; jianying.he@ntnu.no

[2] Porelab, Department of Chemistry, Norwegian University of Science and Technology, NO-7491 Trondheim, Norway; dick.bedeaux@ntnu.no (D.B.); signe.kjelstrup@ntnu.no (S.K.)

* Correspondence: bjorn.a.strom@ntnu.no

Received: 20 July 2020; Accepted: 19 August 2020 ; Published: 27 August 2020

Abstract: Small system properties are known to depend on geometric variables in ways that are insignificant for macroscopic systems. Small system considerations are therefore usually added to the conventional description as needed. This paper presents a thermodynamic analysis of adsorbed films of any size in a systematic and general way within the framework of Hill's nanothermodynamics. Hill showed how to deal with size and shape as variables in a systematic manner. By doing this, the common thermodynamic equations for adsorption are changed. We derived the governing thermodynamic relations characteristic of adsorption in small systems, and point out the important distinctions between these and the corresponding conventional relations for macroscopic systems. We present operational versions of the relations specialized for adsorption of gas on colloid particles, and we applied them to analyze molecular simulation data. As an illustration of their use, we report results for CO_2 adsorbed on graphite spheres. We focus on the spreading pressure, and the entropy and enthalpy of adsorption, and show how the intensive properties are affected by the size of the surface, a feature specific to small systems. The subdivision potential of the film is presented for the first time, as a measure of the film's smallness. For the system chosen, it contributes with a substantial part to the film enthalpy. This work can be considered an extension and application of the nanothermodynamic theory developed by Hill. It provides a foundation for future thermodynamic analyses of size- and shape-dependent adsorbed film systems, alternative to that presented by Gibbs.

Keywords: adsorption; thin film; nanothermodynamics; small-system; size-dependent; thermodynamics; spreading pressure; entropy of adsorption

---

1. Introduction

Adsorption is a central process in nature and in engineering. An important case in nature is adsorption on particles in the atmosphere. It has been recognized for many years that the representation of cloud processes is a significant source of uncertainty in climate models. Many interactions among particles in the atmosphere, clouds and precipitation are relevant for such processes and are the focus of a large area of climate change research [1]. The particles may vary in type, size and shape, and may originate from phenomena such as forest fires and volcanic eruptions, or from human activity such as industry and transportation. They follow the rising air through expansion and cooling, and serve, among other things, as condensation and ice nucleation sites. The initiation of condensation on particles begins with the adsorption or absorption of molecules from the surrounding air. Therefore, the development of more accurate climate models will benefit from a better understanding of these processes for small, curved particles. Although we use herein condensation on particles in the atmosphere as an important application, the focus of this work is on the fundamentals of adsorption on small systems in general.

A substantial amount of research from the 1930s onward on size-effects in thermodynamics may be found, for instance, in the literature on nucleation [2–8]. The majority of the work is based on Gibbs' theory of heterogeneous systems [9]. However, there exists an alternative to the method of Gibbs, developed by Hill in the early 1960s [10]. Hill's method distinguishes itself by the fact that it is a unified and systematic approach to the treatment of all small systems, and does not require the concept of dividing surfaces to be introduced at the outset. The equivalence of Hill's and Gibbs' method in its most general form for the description of curved surfaces has recently been verified [11]. The method is a generalization of macroscopic thermodynamics to small systems, and thus contains conventional thermodynamics as a special case in the macroscopic limit. By generalizing the fundamental differential equations traditionally used in thermodynamics, the whole internal structure of thermodynamic relations for small systems follows naturally by the familiar methods such as Euler integration, which otherwise would not apply for size dependent systems. The analogue of the Gibbs–Duhem equation for small systems is then readily available, and relations describing the size- and shape-dependence of all intensive properties follow. This has motivated us to describe adsorption on small systems using Hill's theory to provide a solid foundation for the present and similar investigations. The aim of the study was thus to present a set of equations for adsorption on a small adsorbent, alternative to existing theories, most prominently Gibbs, and illustrate the set using molecular simulations of CO_2 adsorption on a small sphere of graphite.

In conventional thermodynamics an adsorption system may be described in several ways, one of which considers the system to be the adsorbed phase only. The adsorbent properties are then subtracted from the properties of the surroundings, with the exception of the interaction energy with the adsorbed phase [12]. This view recognizes the asymmetric nature of the system and leads naturally to thermodynamic properties per film molecule. Another approach is to treat the system as a solution of adsorbent and adsorbate [13]. This application is referred to as solution thermodynamics. Though both approaches may be applied, the potential usefulness of one over the other relates to the possible interchange of adsorbent and adsorbate components. In the asymmetric approach, hereafter referred to as adsorption thermodynamics, the focus is on obtaining thermodynamic properties and relations for the adsorbed phase. In the systems typically considered, the properties are categorized as extensive or intensive depending on whether they are proportional to or independent of the size of the film. Therefore, even though the size is necessary in order to completely characterize the film, the nature of the film appears to be independent of it. As a consequence, the differential thermodynamic relations for the adsorbed phase are Euler homogeneous and of first degree in the extensive properties [14]. The relations are therefore directly integrable by the theorem of Euler.

For small systems, the statements above must be modified. The properties normally referred to as intensive are no longer independent of size, so the conventional use of the terms intensive and extensive must now refer to properties that obtain the implied characteristics in the macroscopic limit. The differential thermodynamic relations for a single system are no longer linear homogeneous functions, and can not be directly integrated as before. As the nature of the film now depends on its size and shape, it becomes of interest to systematically investigate the effects of size and shape in a framework that allows for this in a general way. This is possible in the thermodynamic theory of small systems, or nanothermodynamics, as developed by Terrell L. Hill [10], and is the precise reason why we would like to advocate this method. For a more recent text on the method, see Bedeaux et al. [15].

To be able to recognize these differences is important, because it enables us to compare energies of systems that vary in size. Consider two nanoparticles at ambient conditions. One is slightly larger than the other, but they are otherwise identical. The particles do not have bare surfaces; molecules will be adsorbed from the environment—for instance, adsorption of CO_2 on the surfaces of the particles (not adsorption of nano-sized particles on a macroscopic surface). A central question is then: how do the thermodynamic properties of the adsorbed phase differ between the two cases? This is an important question because the adsorbed phase is an interface through which the particle interacts with its environment. Some forms of energy transfer and some forms of interactions are strongly influenced by the presence of adsorbed phases. Not only would size affect the intensive properties, but depending on the constraints on the system, intensive properties may suffice to determine the system size. At length scales where properties typically are size- and shape-dependent, the nanothermodynamics of Hill may help us investigate the interplay between the nature of a system and its size. This is not to say that other approaches may not be successful (see, e.g., [3]), but the corresponding overarching relations (e.g., Maxwell relations) may be less obvious in these. Therefore, the existence of an alternative method may be useful per se.

Our aim was, therefore, to establish operational thermodynamic relations that enable investigations of thin film properties from a nanothermodynamic perspective. Our work has its basis in the work of Hill on nanothermodynamics [10] and papers V [16] and IX [13] in the series on statistical mechanics of adsorption. The relations were applied to molecular simulations of CO_2 adsorption on a spherical graphite-like adsorbent, and to CO_2 adsorption on a generic adsorbent with a strong interaction potential. The purpose of the generic adsorbent case was to observe size effects that were thought to possibly occur when there is significant adsorption on very small particles. The outcome was meant to give a foundation for future, similar developments. The contribution of this work is to clarify and extend the work of Hill on nanothermodynamics and provide a new application.

This paper is organized as follows: In Section 2 we derive general thermodynamic relations for a single-component adsorbent with a single-component adsorbed phase. As the application of Hill's nanothermodynamics is rather limited so far, we considered it necessary to recapitulate the central hypothesis and basis. Readers that are mainly interested in the operational relations and their application may skip this section. Readers that are interested in a comprehensive description of the theory and philosophy of nanothermodynamics are referred to the original work [17] or the recent work [15].

In Section 3 we use the relations to derive operational equations that in turn are applied to the chosen cases. We make simplifying assumptions for the adsorbent, select a reference state typical of adsorption thermodynamics and introduce the condition of equilibrium between the film and the gas. We also derive relations for the dependence of intensive properties on size.

In Section 4 we describe the details of the simulation setup and the methodology for the simulations and the thermodynamic analysis. In Sections 5 and 6 we present and discuss the simulation results, demonstrating the size dependence of a few select thermodynamic properties. We compare the integral free energy per unit area to the differential change in free energy with respect to the change in area. These are properties that are equal in the macroscopic limit. We compare the enthalpy per molecule to the entropy per molecule times temperature, which are also properties that are equal in the macroscopic limit. Finally, in Section 7 we make concluding remarks and propose directions for future work.

2. Nanothermodynamic Framework

In this section, we explain the basic idea of Hill, after considering a common example that motivates his approach; see also Bedeaux et al. [15].

2.1. Small vs. Large Systems

Consider a macroscopic, non-volatile adsorbent of one component with adsorbed adsorbate of one component on the surface, in complete equilibrium with a macroscopic gas adsorbate of the same component. The gas is at temperature T and chemical potential μ, and the external pressure is completely determined by these two variables. The *adsorbent* may then be taken as a thermodynamic system with a fixed number N_A of its component species, in the constant temperature and pressure environment provided by the gas. A large system has the characteristic Gibbs energy $G_A(T, p, N_A) = N_A f(T, p)$. Here f is a function of T and p, but independent of the size N_A and the amount of adsorbed adsorbate N. This implies that contributions of the surface and contributions of the adsorbed adsorbate to the thermodynamic functions of the adsorbent are negligible and that the energy G_A is extensive in N_A. From standard thermodynamics we then have

$$\mu'_A = \left(\frac{\partial G_A}{\partial N_A}\right)_{T,p} = f(T,p) = \frac{G_A}{N_A} \tag{1}$$

where the chemical potential is denoted by a mark to ensure it is not confused with the chemical potential in the general Section 2.

Now consider instead the same system, not large, but small enough for the intensive properties to depend on the size, as measured by the value of N_A. The energy needs to include additional terms related to the size.

The first equality in Equation (1) is still true, but the chemical potential μ'_A is now a different function that depends on the size of the system. Furthermore, G_A is no longer extensive in N_A; thus, the relation $\mu'_A = G_A/N_A$ is no longer true. The relation is re-established in the macroscopic limit, when the terms related to the system size become insignificant. An exact relation, in place of Equation (1), that is valid for the small system may be given using either Gibbs' or Hill's approach. In Hill's approach a new intensive property denoted by the "hat" symbol is introduced to denote the last member of Equation (1); see [17] (p. 1). Here the property is the new chemical potential $\hat{\mu}'_A \equiv G_A/N_A$, but for other environments analogous properties are defined. The introduction of this property helps distinguish between the terms $\mu'_A N_A$ and $\hat{\mu}'_A N_A$ that appear in the integrated forms of the thermodynamic potentials of the macroscopic and small systems, respectively. That these terms are different is a consequence of the fact that the fundamental equations for small systems are not Euler homogeneous. This is the essential difference between macroscopic and small system thermodynamics. Both μ'_A and $\hat{\mu}'_A$ for small systems are functions of N_A in addition to T and p, and are therefore different from the macroscopic chemical potential. In the macroscopic limit μ'_A and $\hat{\mu}'_A$ both become equal to the macroscopic chemical potential.

When the environment variables include multiple extensive variables, e.g., T, p, N_1, N_2, ..., new intensive properties conjugate to the extensive variables are not defined. Instead, the energy term is referred to as X, and depends on the environment. This notation still distinguishes it from the respective terms, e.g., $X \neq \mu_1 N_1 + \mu_2 N_2 + \ldots$, in the integrated form of the thermodynamic potential for the macroscopic system, but it does not indicate what type of energy it is per se.

2.2. Hill's Extension

The example above shows how properties of small systems may depend on the system size. A relevant theory must therefore allow for variations in size. The theory needs, furthermore, to produce thermodynamic functions and relationships for a single small system. Hill was able to extend large scale thermodynamics to include small systems. In the macroscopic limit his theory becomes identical to the standard thermodynamic equations.

Macroscopic thermodynamics can be applied to a large sample of small systems, such as a macromolecular solution where the small system is the solute macromolecule. In this description we may change the number of small systems, but we cannot change the size of the small system itself. To have a solid foundation for the theory and at the same time allow for variations in size, Hill used the macroscopic thermodynamics of an ensemble of independent small systems as his starting point and introduced size-determining properties as variable parameters. An ensemble is a large collection of systems, where each system replicates the thermodynamic state and environment of the actual thermodynamic system [18]. A member of the ensemble may thus be referred to as a replica.

The system of practical interest in this work is an adsorbent consisting of a single component of quantity N_A with an adsorbed phase consisting of a single component of quantity N. The system is at temperature T, pressure p and adsorbed component chemical potential μ, and is completely characterized by T, p, μ and N_A. A real system of this type will always exist in the presence of adsorbate molecules—for instance, as free molecules in a gas. The distinction between adsorbed and free molecules is then somewhat arbitrary. There is nothing within the framework of thermodynamics that provides a unique definition of the distinction, so this information must come from elsewhere, such as from experiment or from a theoretical model. In order to stay as general as possible, we make no assumptions in this regard for the time being; cf. Section 4.

In the present case, the method of Hill considers an ensemble of $\mathcal{N}$ independent small systems at temperature T, pressure p and adsorbed component chemical potential μ. All systems have the same assigned amount of adsorbent component N_A. The ensemble, with total properties denoted by subscript t, can now be characterized by the entropy S_t, volume V_t, amount of adsorbed component N_t and amount of adsorbent component $\mathcal{N} N_A$. We allow for an independent variation in the number of small systems $\mathcal{N}$ and also in the size of the small systems, as given by N_A. The characteristic function for the ensemble in terms of the set of independent variables $S_t, V_t, N_t, N_A, \mathcal{N}$ is the internal energy U_t given by

$$dU_t = T\,dS_t - p\,dV_t + \mu\,dN_t + \mu_A \mathcal{N}\,dN_A + X\,d\mathcal{N} \tag{2}$$

where X, the replica energy, may from Equation (2) be formally defined by

$$X \equiv \left(\frac{\partial U_t}{\partial \mathcal{N}}\right)_{S_t, V_t, N_t, N_A} \tag{3}$$

The function X describes how a change $d\mathcal{N}$ in the number of small systems will change the internal energy of the entire ensemble through $X\,d\mathcal{N}$. From a physical perspective $X\,d\mathcal{N}$ refers to the process of adding systems to the ensemble, each with the same amount of adsorbent component N_A, while keeping $S_t = \mathcal{N}S$, $V_t = \mathcal{N}\bar{V}$, and $N_t = \mathcal{N}\bar{N}$ constant. This process explains why X may be referred to as the replica energy. For the distribution of small systems over the possible states, this means that the entropy S, the mean volume $\bar{V}$ and the mean amount of adsorbed component $\bar{N}$ must decrease. In the process, the ensemble gains the amount $N_A\,d\mathcal{N}$ of adsorbent component, as opposed to the amount $\mathcal{N}dN_A$, while the properties S_t, V_t and N_t must be redistributed across the new number of systems $\mathcal{N} + d\mathcal{N}$. The term $X\,d\mathcal{N}$ is of the same type as the chemical potential term $\mu_A \mathcal{N}\,dN_A$. However, μ_A refers to the addition of a differential amount dN_A of adsorbent component for each replica, $\mathcal{N}\,dN_A$ in total, while X refers to the addition of a single system, or an integral amount of adsorbent component N_A. We see this also by looking at the last two terms in Equation (2) which may be rewritten as $\mu_A \mathcal{N}\,dN_A + X\,d\mathcal{N} = \mu_A\,d(\mathcal{N}N_A) + (X - \mu_A N_A)\,d\mathcal{N} = \mu_A\,dN_{t,A} + (X - \mu_A N_A)\,d\mathcal{N}$. In the macroscopic limit the ensemble energy is completely characterized by S_t, V_t, N_t and $N_{t,A}$ and the term $(X - \mu_A N_A)\,d\mathcal{N}$ is zero. It follows that in the macroscopic limit $X \to \mu_A N_A$. This motivates the definition of the new function $\hat{\mu}_A$ by $\hat{\mu}_A N_A \equiv X$ such that $\hat{\mu}_A \to \mu_A$ in the macroscopic limit. Both $\hat{\mu}_A$ and μ_A differ from the macroscopic chemical potential at the same T and p.

The small system can therefore be said to differ from a large one, by the thermodynamic function $\mathcal{E}$, where $\mathcal{E} \equiv (\hat{\mu}_A - \mu_A)N_A$. This function, which can be viewed as a correction from macroscopic thermodynamics, has been called the subdivision potential, because the process of subdividing the ensemble into a larger number of smaller systems, while keeping S_t, V_t, N_t and $N_{t,A}$ constant, requires the energy $\mathcal{E}\,d\mathcal{N}$. It has recently been shown to contain interesting size-scaling laws [15,19,20]. The replica energy, the subdivision potential and the scaling laws are particular for the variables that control the ensemble. The presented results refer to an osmotic ensemble, with environment variables T, p, μ and N_A.

Next come the thermodynamic functions that follow from Equation (2). By integrating Equation (2) at constant T, p, μ, N_A; defining $U_t \equiv \mathcal{N}\bar{U}$, $S_t \equiv \mathcal{N}S$, $V_t \equiv \mathcal{N}\bar{V}$ and $N_t \equiv \mathcal{N}\bar{N}$; and dividing by $\mathcal{N}$ we obtain the mean internal energy of a single small system given by

$$\bar{U} = TS - p\bar{V} + \mu\bar{N} + \hat{\mu}_A N_A \tag{4}$$

By substituting the definitions for $\bar{U}$, S, $\bar{V}$ and $\bar{N}$ in Equation (2) and eliminating $\hat{\mu}_A N_A$ by Equation (4), it follows that

$$d\bar{U} = T\,dS - p\,d\bar{V} + \mu\,d\bar{N} + \mu_A\,dN_A \tag{5}$$

The differential change in internal energy for a single small system given by Equation (5) has the same form as for a macroscopic system, with the important distinction that the intensive properties are functions of the size of the system. The energy $\bar{U}$ is therefore not a linear homogeneous function of S, $\bar{V}$, $\bar{N}$ and N_A. Therefore we can not obtain Equation (4) by Euler integration of Equation (5). This can be seen clearly if we rewrite Equation (4) as

$$\bar{U} = TS - p\bar{V} + \mu\bar{N} + \mu_A N_A + \mathcal{E} \tag{6}$$

where we have used the previous definition $\mathcal{E} \equiv (\hat{\mu}_A - \mu_A)N_A$. The equation is the same as we would expect for a macroscopic system except for the extra term $\mathcal{E}$. This general feature of small systems has important implications for the Clausius–Clapeyron type equation (Equation (36)), and for the analogue for a small system of the Gibbs adsorption isotherm; see the discussion below in relation to Equation (48). By differentiating Equation (4) and subtracting Equation (5), we obtain

$$d(\hat{\mu}_A N_A) = -S\,dT + \bar{V}\,dp - \bar{N}\,d\mu + \mu_A\,dN_A \tag{7}$$

This is the function characteristic of a single system of the ensemble for the independent variables T, p, μ, N_A. It is of particular interest operationally because the independent variables are the environment variables. By subtracting $d(\mu_A N_A)$ from Equation (7) we have

$$d\mathcal{E} = -S\,dT + \bar{V}\,dp - \bar{N}\,d\mu - N_A\,d\mu_A \tag{8}$$

This equation shows that a small system has one additional independent variable compared to a macroscopic system. In the macroscopic limit, $\mathcal{E} = 0$ and Equation (8) becomes the Gibbs–Duhem equation, a relation between the intensive variables T, p, μ and μ_A such that only three of them are independent. From Equations (7) and (8) it follows that

$$d\hat{\mu}_A = -\frac{S}{N_A}\,dT + \frac{\bar{V}}{N_A}\,dp - \frac{\bar{N}}{N_A}\,d\mu - \frac{\mathcal{E}}{N_A^2}\,dN_A \tag{9}$$

$$d\mu_A = -\left(\frac{\partial S}{\partial N_A}\right)_{T,p,\mu} dT + \left(\frac{\partial \bar{V}}{\partial N_A}\right)_{T,p,\mu} dp - \left(\frac{\partial \bar{N}}{\partial N_A}\right)_{T,p,\mu} d\mu - \frac{1}{N_A}\left(\frac{\partial \mathcal{E}}{\partial N_A}\right)_{T,p,\mu} dN_A \tag{10}$$

The expression for $d\mu_A$ is obtained by considering μ_A a function of T, p, μ and N_A, and writing the general expression for the differential. The first three differential coefficients, i.e., in T, p and μ are obtained by Maxwell relations from Equation (7). By substituting the general expression for $d\mu_A$ in Equation (8) and setting T, p, μ constant, we may solve for the last differential coefficient. Equations (9) and (10) show the integral-differential relation between $\hat{\mu}_A$ and μ_A.

We define the Gibbs energy G and the enthalpy H by

$$\bar{G} \equiv \bar{U} - TS + p\bar{V} = \mu\bar{N} + \mu_A N_A + \mathcal{E} = \mu\bar{N} + \hat{\mu}_A N_A \tag{11}$$

$$\bar{H} \equiv \bar{G} + TS = TS + \mu\bar{N} + \hat{\mu}_A N_A \tag{12}$$

This concludes the theoretical basis needed to describe adsorption on a small adsorbent. We proceed to make the relations operational.

3. Operational Relations

The thermodynamic system was defined above as the adsorbent plus the adsorbed film. We are interested in the properties of the film, and how they vary with the size of the adsorbent. The aim is to be able to plot adsorption isotherms, spreading pressure and corresponding film entropy and enthalpy.

3.1. The Reference State

As is the usual procedure in adsorption thermodynamics, we define the reference state as the quantity, N_A, of pure adsorbent, with volume $\bar{V}_{0A}$ at external pressure, p, and temperature T. By pure adsorbent we mean the adsorbent with a clean surface in the absence of adsorption, $N \rightarrow 0$. The quantities T, p and N_A were defined above; see Section 2. The properties of the film can then be defined as the properties of the total system relative to the reference. In practice, we subtract properties of the pure adsorbent from the total system, while keeping the interaction energy with the film molecules. The equivalents of Equations (4), (5) and (7) for the reference are given by

$$\bar{U}_{0A} = TS_{0A} - p\bar{V}_{0A} + \hat{\mu}_{0A}N_A \tag{13}$$

$$d\bar{U} = T\,dS_{0A} - p\,d\bar{V}_{0A} + \mu_{0A}\,dN_A \tag{14}$$

$$d(\hat{\mu}_{0A}N_A) = -S_{0A}\,dT + \bar{V}_{0A}\,dp + \mu_{0A}\,dN_A \tag{15}$$

By subtracting Equation (13) from Equation (4), Equation (14) from Equation (5) and Equation (15) from Equation (7), we obtain the equations for the film.

$$\bar{U}_s = TS_s - p\bar{V}_s + \mu\bar{N}_s - \hat{\Phi}N_A \tag{16}$$

$$d\bar{U}_s = T\,dS_s - p\,d\bar{V}_s + \mu\,d\bar{N} - \Phi\,dN_A \tag{17}$$

$$d(\hat{\Phi}N_A) = S_s\,dT - \bar{V}_s\,dp + \bar{N}_s\,d\mu + \Phi\,dN_A \tag{18}$$

where Equation (18) is the characteristic equation for the film in the given environment. Superscript F denotes the film properties defined by $S_s \equiv S - S_{0A}$, $\bar{V}_s \equiv \bar{V} - \bar{V}_{0A}$, $\bar{N}_s \equiv \bar{N}$, $\hat{\Phi} \equiv \hat{\mu}_{0A} - \hat{\mu}_A$ and $\Phi \equiv \mu_{0A} - \mu_A$. The property $\bar{N}_s$ is defined for consistency in notation because $\bar{N}$ is already the amount of adsorbed component only.

3.2. Size-Dependent Thermodynamic Properties

The analysis continues from here under the approximations that the adsorbent is unaffected by the adsorption, and that its volume, shape and structure are independent of temperature and pressure. The adsorbent then functions only as an external field acting on the adsorbed phase. A more general approach is required for adsorbents that evaporate/dissolve; adsorbents whose structure is affected by the adsorption; and cases where there is no unambiguous definition of the adsorbent's surface area [13]. However, for the calculations done here, we do not require that level of generality. In general the adsorbent volume may be considered to be a function of T, p and N_A. However, under the current approximations the adsorbent thermal expansion and compressibility are negligible. It follows that the adsorbent volume and surface area are functions of N_A only. For a constant spherical shape, there is now only one independent variable among the adsorbent volume, surface area and N_A. The adsorbent surface area Ω is a natural choice when we want to describe the properties of the film only. Eliminating N_A in Equations (17) and (18) by substituting $dN_A = (dN_A/d\Omega)\,d\Omega$, we have

$$d\bar{U}_s = T\,dS_s - p\,d\bar{V}_s + \mu\,d\bar{N} - \varphi\,d\Omega \tag{19}$$

$$d(\hat{\varphi}\Omega) = S_s\,dT - \bar{V}_s\,dp + \bar{N}_s\,d\mu + \varphi\,d\Omega \tag{20}$$

where we have defined $\varphi \equiv \Phi(dN_A/d\Omega)$ and $\hat{\varphi} \equiv \hat{\Phi}(N_A/\Omega)$. The property φ is the usual spreading pressure in adsorption thermodynamics [12]. The property $\hat{\varphi}$ is related to the subdivision potential $\mathcal{E}_s$ by $\mathcal{E}_s = (\varphi - \hat{\varphi})\Omega$. The relation can be taken as a definition of the subdivision potential. The equivalents of Equations (9) and (10) are

$$d\hat{\varphi} = \frac{S_s}{\Omega}\,dT - \frac{\bar{V}_s}{\Omega}\,dp + \frac{\bar{N}_s}{\Omega}\,d\mu + \frac{\mathcal{E}_s}{\Omega^2}\,d\Omega \tag{21}$$

$$d\varphi = \left(\frac{\partial S_s}{\partial\Omega}\right)_{T,p,\mu}dT - \left(\frac{\partial\bar{V}_s}{\partial\Omega}\right)_{T,p,\mu}dp + \left(\frac{\partial\bar{N}_s}{\partial\Omega}\right)_{T,p,\mu}d\mu + \frac{1}{\Omega}\left(\frac{\partial\mathcal{E}_s}{\partial\Omega}\right)_{T,p,\mu}d\Omega \tag{22}$$

Due to the integral forms of the coefficients in Equation (21) and the differential forms of the corresponding coefficients in Equation (22), we refer to $\hat{\varphi}$ as the integral spreading pressure and to φ as the differential spreading pressure. For Equations (11) and (12) relative to the reference, we see that

$$\bar{G} - \bar{G}_{0A} = \mu\bar{N}_s - \hat{\varphi}\Omega = \mu\bar{N}_s - \varphi\Omega + \mathcal{E}_s \tag{23}$$

$$\bar{H} - \bar{H}_{0A} = TS_s + \mu\bar{N}_s - \hat{\varphi}\Omega = TS_s + \mu\bar{N}_s - \varphi\Omega + \mathcal{E}_s \tag{24}$$

Following the usual procedure in adsorption thermodynamics, the term $\varphi\Omega$ is recognized as the analogue of pV for ordinary three-dimensional thermodynamics. We therefore define the film functions $\bar{G}_s$ and $\bar{H}_s$ by

$$\bar{G}_s \equiv \bar{G} - \bar{G}_{0A} + \varphi\Omega = \mu\bar{N}_s + \mathcal{E}_s \tag{25}$$

$$\bar{H}_s \equiv \bar{G}_s + TS_s = TS_s + \mu\bar{N}_s + \mathcal{E}_s \tag{26}$$

This ensures the relation between $\bar{H}_s$ and S_s given in Equation (39) when the system is in equilibrium with the gas. The relation in Equation (39) becomes the one usually encountered in adsorption thermodynamics when the system is macroscopic and $\mathcal{E}_s = 0$. By Equations (21) and (22) and the relation $\mathcal{E}_s = (\varphi - \hat{\varphi})\Omega$ we may write $d\mathcal{E}_s$ in terms of the environment variables T, p, μ, Ω as

$$d\mathcal{E}_s = \Omega^2\left(\frac{\partial S_s/\Omega}{\partial\Omega}\right)_{T,p,\mu}dT - \Omega^2\left(\frac{\partial\bar{V}_s/\Omega}{\partial\Omega}\right)_{T,p,\mu}dp + \Omega^2\left(\frac{\partial\bar{N}_s/\Omega}{\partial\Omega}\right)_{T,p,\mu}d\mu + \Omega\left(\frac{\partial\varphi}{\partial\Omega}\right)_{T,p,\mu}d\Omega \tag{27}$$

The differential coefficients of the type $\partial(Y/\Omega)/\partial\Omega$ in Equation (27), where $Y = S_s, \bar{V}_s, \bar{N}_s$, may be expanded as

$$\left(\frac{\partial Y/\Omega}{\partial\Omega}\right)_{T,p,\mu} = \frac{1}{\Omega}\left[\left(\frac{\partial Y}{\partial\Omega}\right)_{T,p,\mu} - \frac{Y}{\Omega}\right] \tag{28}$$

This emphasizes the distinction between differential and integral quantities for small systems, one of the key points of nanothermodynamics. In the macroscopic limit, linear homogeneous relations of the type $\partial Y/\partial\Omega = Y/\Omega$ are again true and the bracket term vanishes. It follows directly from Equation (27) that the effects of size on the intensive variables S_s/Ω, $\bar{V}_s/\Omega$, $\bar{N}_s/\Omega$ and φ may be related to $\mathcal{E}_s$ and changes in $\mathcal{E}_s$ by

$$\left(\frac{\partial S_s/\Omega}{\partial\Omega}\right)_{T,p,\mu} = \frac{1}{\Omega^2}\left(\frac{\partial\mathcal{E}_s}{\partial T}\right)_{p,\mu,\Omega} \tag{29}$$

$$\left(\frac{\partial\bar{V}_s/\Omega}{\partial\Omega}\right)_{T,p,\mu} = -\frac{1}{\Omega^2}\left(\frac{\partial\mathcal{E}_s}{\partial p}\right)_{T,\mu,\Omega} \tag{30}$$

$$\left(\frac{\partial\bar{N}_s/\Omega}{\partial\Omega}\right)_{T,p,\mu} = \frac{1}{\Omega^2}\left(\frac{\partial\mathcal{E}_s}{\partial\mu}\right)_{T,p,\Omega} \tag{31}$$

$$\left(\frac{\partial\varphi}{\partial\Omega}\right)_{T,p,\mu} = \frac{1}{\Omega}\left(\frac{\partial\mathcal{E}_s}{\partial\Omega}\right)_{T,p,\mu} \tag{32}$$

From Equation (21) we have

$$\left(\frac{\partial \hat{\varphi}}{\partial \Omega}\right)_{T,p,\mu} = \frac{\mathcal{E}_s}{\Omega^2} \tag{33}$$

These are the overarching relations inherent in Hill's formalism that we next may take advantage of in descriptions of further properties of the film. They show the central role of the subdivision potential, which can serve as a direct measure of the system smallness, as we shall see below, e.g., in Equation (37). The relations are not directly obtainable in Gibbs' treatment of adsorption. From Equation (26) using Equations (29) and (31) it follows that

$$\left(\frac{\partial H_s/\Omega}{\partial \Omega}\right)_{T,p,\mu} = \frac{T}{\Omega^2}\left(\frac{\partial \mathcal{E}_s}{\partial T}\right)_{p,\mu,\Omega} + \frac{\mu}{\Omega^2}\left(\frac{\partial \mathcal{E}_s}{\partial \mu}\right)_{T,p,\Omega} + \left(\frac{\partial \mathcal{E}_s/\Omega}{\partial \Omega}\right)_{T,p,\mu} \tag{34}$$

We now consider the special case where the system is in equilibrium with a macroscopic gas at T, p. The chemical potential μ is then equal to the gas chemical potential μ_G. It follows that μ is completely determined by T and p and that a change $d\mu$ is given by $d\mu = -s_G\,dT + v_G\,dp$, where s_G and v_G are the entropy and volume per gas molecule. Equation (21) becomes

$$d\hat{\varphi} = (s_s - s_G)\bar{\Gamma}_s\,dT - (\bar{v}_s - v_G)\bar{\Gamma}_s\,dp + \mathcal{E}_s/\Omega^2\,d\Omega \tag{35}$$

where $s_s \equiv S_s/\bar{N}_s$, $\bar{v}_s \equiv \bar{V}_s/\bar{N}_s$, and $\bar{\Gamma}_s \equiv \bar{N}_s/\Omega$. It follows that the entropy per film molecule relative to the entropy per gas molecule at the same conditions is given by

$$s_s - s_G = (\bar{v}_s - v_G)\left(\frac{\partial p}{\partial T}\right)_{\hat{\varphi},\Omega} \tag{36}$$

This equation together with Equation (46) is applied below to calculate the entropy per film molecule relative to the gas. The relation between the chemical potential of the film and the enthalpy follows from Equation (26), and is given by

$$\mu = \bar{h}_s - Ts_s - \mathcal{E}_s/\bar{N}_s \tag{37}$$

where $\bar{h}_s = \bar{H}_s/\bar{N}_s$. From the equilibrium condition $\mu = \mu_G$ it follows that

$$\bar{h}_s - Ts_s - \mathcal{E}_s/\bar{N}_s = h_G - Ts_G \tag{38}$$

$$\bar{h}_s - h_G = T(s_s - s_G) + \mathcal{E}_s/\bar{N}_s \tag{39}$$

This equation was used to calculate the enthalpy per film molecule relative to the gas in simulations.

We can write the equations that relate the effects of size on intensive variables to $\mathcal{E}_s$ in simpler form. Equation (27) becomes

$$d\mathcal{E}_s = \Omega^2\left[\frac{\partial(s_s - s_G)\bar{\Gamma}_s}{\partial \Omega}\right]_{T,p} dT - \Omega^2\left[\frac{\partial(\bar{v}_s - v_G)\bar{\Gamma}_s}{\partial \Omega}\right]_{T,p} dp + \Omega\left(\frac{\partial \varphi}{\partial \Omega}\right)_{T,p} d\Omega \tag{40}$$

The operational equivalents of Equations (29)–(32), now follow directly from Equation (40):

$$\left[\frac{\partial(s_s - s_G)\bar{\Gamma}_s}{\partial \Omega}\right]_{T,p} = \frac{1}{\Omega^2}\left(\frac{\partial \mathcal{E}_s}{\partial T}\right)_{p,\Omega} \tag{41}$$

$$\left[\frac{\partial(\bar{v}_s - v_G)\bar{\Gamma}_s}{\partial \Omega}\right]_{T,p} = -\frac{1}{\Omega^2}\left(\frac{\partial \mathcal{E}_s}{\partial p}\right)_{T,\Omega} \tag{42}$$

$$\left(\frac{\partial \varphi}{\partial \Omega}\right)_{T,p} = \frac{1}{\Omega}\left(\frac{\partial \mathcal{E}_s}{\partial \Omega}\right)_{T,p} \tag{43}$$

From Equation (35) we have the operational equivalent of Equation (33):

$$\left(\frac{\partial \hat{\varphi}}{\partial \Omega}\right)_{T,p} = \frac{\mathcal{E}_s}{\Omega^2} \tag{44}$$

From Equation (39) using Equation (41), we have the operational equivalent of Equation (34):

$$\left[\frac{\partial (h_s - h_G)\bar{\Gamma}_s}{\partial \Omega}\right]_{T,p} = \frac{T}{\Omega^2}\left(\frac{\partial \mathcal{E}_s}{\partial T}\right)_{p,\Omega} + \left(\frac{\partial \mathcal{E}_s/\Omega}{\partial \Omega}\right)_{T,p} \tag{45}$$

3.3. Analogue of the Gibbs Adsorption Isotherm

We can integrate Equation (35) at constant T and Ω to obtain $\hat{\varphi}$ if we have the adsorption isotherm for a given adsorbent size. We then have

$$\hat{\varphi} = \int_0^p \bar{\Gamma}_s(v_G - \bar{v}_s)\,dp, \quad (T, \Omega \text{ const.}) \tag{46}$$

The choice of reference system used to define the film properties can be motivated by this operational equation. As the gas pressure decreases towards zero, the system approaches the reference system. Therefore, $\hat{\varphi}$ vanishes at the lower integration limit. The role of the subdivision potential is no longer visible in the end formula, Equation (46), but as has been seen above, the property can be regarded as an expression of a certain internal structure that must be obeyed.

In standard thermodynamics the well known Gibbs adsorption isotherm may be obtained from the analogue for a surface phase of the Gibbs–Duhem relation at constant temperature [12]. The Gibbs–Duhem relation may be obtained by a standard procedure involving Euler integration of the expression for the differential change in internal energy. When the pure adsorbent is used as the reference state, and the criterion for the distinction between free and adsorbed adsorbate is such that $V_s = 0$, the analogue of the Gibbs adsorption isotherm is given by Equation (1002,8) in [12], or Equation (15) in [21]. At equilibrium, assuming the gas is an ideal gas, we have $d\mu = d\mu_G = v_G\,dp \approx kT\,d\ln p$. It follows that the integrated form of the Gibbs adsorption isotherm is given by Equation (29) in [13], or Equation (19) in [21].

The important point here is that direct Euler integration of Equation (19) is not possible. Therefore, the ensemble procedure is used to obtain the analogue of the Gibbs–Duhem relation for a small system, given by

$$d\mathcal{E}_s = d[(\varphi - \hat{\varphi})\Omega] = -S_s\,dT + \bar{V}_s\,dp - \bar{N}_s\,d\mu + \Omega\,d\varphi \tag{47}$$

which in integrated form at constant T and Ω gives Equation (46). Equation (47) further shows the central role of the subdivision potential. Assuming that the gas is an ideal gas, and that the criterion adopted for the distinction between free and adsorbed adsorbate is such that $\bar{V}_s = 0$, the analogue of the Gibbs adsorption isotherm for a small system follows from Equation (46), and is given by

$$\hat{\varphi} = kT \int_0^p \bar{\Gamma}_s\,d\ln p, \quad (T, \Omega \text{ const.}) \tag{48}$$

In the macroscopic limit the dependence of the intensive properties on the size of the system becomes negligible, and we have $\hat{\varphi} = \varphi$ in Equation (48).

It is interesting to compare the relation between φ and $\hat{\varphi}$ to the relation between μ'_A and $\hat{\mu}'_A$; cf. Section 2. Let the adsorbent on which the film is formed be macroscopic such that the film may be considered flat on a scale that is small compared to the size of the adsorbent, but still large compared to the film thickness. The film is then a thermodynamic system characterized by T, p and Ω. Suppose the film has the energy $\hat{\varphi}\Omega = f(T, p)\Omega$. The area Ω is then necessary in order to completely characterize the film, but the nature of the film is independent of it. It follows from this expression and Equation (20), with $\mu = \mu(T, p)$, that

$$\varphi = \left(\frac{\partial \hat{\varphi}\Omega}{\partial \Omega}\right)_{T,p} = f(T, p) = \hat{\varphi} \tag{49}$$

Now consider instead that the system is not macroscopic. The energy may then be alternatively described by including additional terms related to the size of the system. For instance, suppose that the energy is given by

$$\hat{\varphi}\Omega = f(T, p)\Omega + g(T, p, \Omega) \tag{50}$$

We would then have

$$\varphi = \left(\frac{\partial \hat{\varphi}\Omega}{\partial \Omega}\right)_{T,p} = f + \left(\frac{\partial g}{\partial \Omega}\right)_{T,p} \tag{51}$$

$$\frac{\mathcal{E}_s}{\Omega} = \varphi - \hat{\varphi} = \left(\frac{\partial g}{\partial \Omega}\right)_{T,p} - \frac{g}{\Omega} \tag{52}$$

We see that while the first equality in Equation (49) still holds, the second one does not. Therefore, in order to obtain generalized thermodynamic equations that are valid for the small system represented by Equation (50) and that become the conventional thermodynamic equations in the macroscopic limit, we must, in addition to the regular spreading pressure φ, define the integral spreading pressure $\hat{\varphi}$. The two functions become equal in the macroscopic limit. Then the energy $\mathcal{E}_s \, d\mathcal{N}$ required to subdivide the ensemble into a larger number of smaller systems is negligible. It then follows from Equation (52) that the difference between the differential property $(\partial g/\partial \Omega)_{T,p}$ and integral property g/Ω is negligible.

4. Methodology

4.1. Simulation Techniques

Molecular simulations were done using the open source software LAMMPS (version 20 August 2019) [22] with the grand canonical Monte Carlo simulation technique. A simulation was run for each thermodynamic state of the film characterized by T, p and Ω. All simulations were bounded by a cubic simulation box with periodic boundary conditions. The size of the box varied depending on the state, but was fixed for any given state. As an example, for all states of the graphite adsorbent system consistent with $T = 1.08$ and $\Omega \approx 82$, the side length of the box was approximately 15 in reduced units. The gas pressure and density were sampled sufficiently far away from the adsorbent surface for the gas to obtain bulk properties.

The interaction between the gas molecules was described by the Mie potential [23].

$$u_G(r) = B\epsilon_G \left[\left(\frac{\sigma_G}{r}\right)^{\gamma_r} - \left(\frac{\sigma_G}{r}\right)^{\gamma_a}\right], \quad B = \left(\frac{\gamma_r}{\gamma_r - \gamma_a}\right)\left(\frac{\gamma_r}{\gamma_a}\right)^{\left(\frac{\gamma_a}{\gamma_r - \gamma_a}\right)} \tag{53}$$

where ϵ_G is the energy parameter of the interaction, σ_G is the length parameter of the interaction, γ_r is the repulsive exponent and γ_a is the attractive exponent. The parameters were taken from [24] to represent single site coarse-grained CO_2 molecules. For the unit of energy we used the strength of the Mie interaction 361.69 kJ, where k is the Boltzmann constant. For the unit of length we used the diameter of the CO_2 Mie segment 3.741 Å. For the unit of mass, we used the molecular mass of CO_2. We set the parameters as follows: (1) The number of Mie segments $m_s = 1$, (2) $\sigma_G = 1$ by the definition of units, (3) $\epsilon_G = 1$ by the definition of units, (4) $\gamma_r = 23.0$, (5) $\gamma_a = 6.66$ and (6) the potential cut-off distance $r_{c,mie} = 4.0$.

The interaction between the adsorbent functioning as an external field and a gas molecule, was represented by a spherical colloid potential located at the box center. The expression for this potential follows by integration of the pairwise interactions between a gas molecule and the adsorbent constituent atoms over the adsorbent volume. The interaction between a gas molecule and the adsorbent constituent atom was given by the standard Lennard–Jones 12-6 potential

$$u(r') = 4\epsilon \left[\left(\frac{\sigma}{r'} \right)^{12} - \left(\frac{\sigma}{r'} \right)^{6} \right] \tag{54}$$

where $u(r')$ is the interaction energy between an adsorbent atom and a gas molecule, r' is the center to center distance between them, ϵ is the energy parameter of the interaction and σ is the length parameter of the interaction. By integrating Equation (54) over the spherical adsorbent, we have

$$\mathcal{U}(r) = \frac{16\pi\epsilon\rho_A\sigma^3}{3} \left[\frac{(15R^3r^6 + 63R^5r^4 + 45R^7r^2 + 5R^9)\sigma^9}{15(r^2 - R^2)^9} - \frac{R^3\sigma^3}{(r^2 - R^2)^3} \right], \quad r > R \tag{55}$$

where $\mathcal{U}(r)$ is the interaction energy between the adsorbent and a gas molecule, r is the center to center distance between them, ρ_A is the adsorbent number density and R is the adsorbent radius. The cut-off distance r_c for the potential was defined by the relation $\mathcal{U}(r_c)/\min(\mathcal{U}) = 5.5 \times 10^{-3}$, where $\min(\mathcal{U})$ is the potential minimum. The value of r_c, satisfying the relation, was approximated by $r_c = 1.23R + 3.0$ for each adsorbent radius. We used the conventional definition of the Hamaker constant $\mathcal{A}_{12} \equiv \pi^2 C\rho_1\rho_2$, where C is the coefficient of the dispersion energy $u_{disp}(r') = -C/(r')^6$. For a potential of the type in Equation (54) it follows that $C = 4\epsilon_{12}\sigma_{12}^6$ and

$$\mathcal{A}_{12} = 4\pi^2\epsilon_{12}\rho_1\rho_2\sigma_{12}^6 \tag{56}$$

Using Equation (56), and the mixing rules $\sigma = (\sigma_G + \sigma_A)/2$ and $\epsilon = \sqrt{\epsilon_G\epsilon_A}$, we may calculate $\epsilon\rho_A\sigma^3$ in terms of σ_G, σ_A, ϵ_G, and $\mathcal{A}_A$ for use in Equation (55). This assumed the adsorbent had the same density as the bodies for which $\mathcal{A}_A$ was measured. Here σ_A and ϵ_A are the length and energy parameters of the dispersion energy of two interacting adsorbent atoms, analogous to σ and ϵ, and $\mathcal{A}_A$ is the Hamaker constant for the interaction between two graphite bodies. We take $\sigma_A = 3.4$ Å for the carbon atom–atom interaction, and $\mathcal{A}_A = 4.7 \times 10^{-19}$ J for the Hamaker constant [25]. It follows that $\epsilon\rho_A\sigma^3 = [\sqrt{\epsilon_G\mathcal{A}_A}/(16\pi)][(\sigma_G + \sigma_A)/\sigma_A]^3 \approx 1.79$ in reduced units for the graphite adsorbent system. For the generic adsorbent system we used $\epsilon\rho_A\sigma^3 = 11.0$, and the cut-off was set to the fixed value $r_c = 7$.

In order to use the thermodynamic relations with the simulation data, we need to be able to distinguish between adsorbed and free molecules. We therefore define the location of a spherical mathematical dividing surface at a radial distance a from the box center by the criterion $\mathcal{U}(a) = 0$. We then define the adsorbent volume by $V_A \equiv (4/3)\pi a^3$, and the number of adsorbed molecules by $N \equiv N_B - \rho_G(V_B - V_A)$, where N_B is the number of adsorbate molecules in the box, ρ_G is the gas density and V_B is the box volume. This is a definition based on the concept of surface excess properties introduced by Gibbs [9]. It was easy to apply to the simulations because we had access to the total quantity of the adsorbing component, the gas density and the volume.

4.2. Data Reduction

Plots of the integral spreading pressure $\hat{\varphi}(\Omega; T, p)$ were first obtained from the adsorption isotherm simulation data of the type $\bar{\Gamma}_s(p; T, \Omega)$, one for each area. The following prescription was used to find the desired properties:

- The isotherms were interpolated as described below and integrated using Equation (46), approximating the gas as ideal such that $v_G = kT/p$.
- Each of the resulting functions $\hat{\varphi}(p; T, \Omega)$, one for each area, were then evaluated at the desired pressure to give the final curve.
- The curve for the differential spreading pressure, φ, was obtained from the functions $\hat{\varphi}(p; T, \Omega)$ for all areas, from the relation $\varphi = \hat{\varphi} + \Omega(\partial\hat{\varphi}/\partial\Omega)_{T,p}$, which follows from Equation (35) at constant

T, p and the definition $\mathcal{E}_s \equiv (\varphi - \hat{\varphi})/\Omega$. The derivative $\Omega(\partial\hat{\varphi}/\partial\Omega)_{T,p}$ was approximated by $\Omega(\Delta\hat{\varphi}/\Delta\Omega)_{T,p}$ which was calculated from two functions $\hat{\varphi}(p; T, \Omega_1)$ and $\hat{\varphi}(p; T, \Omega_2)$ for values of Ω_1 and Ω_2 not too far apart.

- The curve for $\mathcal{E}_s/\Omega$ was obtained as the difference $\varphi - \hat{\varphi}$.
- The curves for the entropy and enthalpy were obtained by Equations (36) and (39), respectively. This required the use of Equation (46) first, so that we knew $\hat{\varphi}(p; T, \Omega)$. The derivative term was approximated by $-(kT/p_1)(\partial p/\partial T)_{\hat{\varphi},\Omega} \approx -(kT/p_1)(\Delta p/\Delta T)$ which was calculated from two functions $p(\hat{\varphi}; T_1, \Omega)$ and $p(\hat{\varphi}; T_2, \Omega)$ for values of T_1 and T_2 not too far apart. The two temperatures used were $T_1 = 1.080$ and $T_2 = 1.165$. The functions were obtained by inversion of $\hat{\varphi}(p; T_1, \Omega)$ and $\hat{\varphi}(p; T_2, \Omega)$.

For all state properties of which the approximate value was calculated from two states, i.e., from expressions containing finite difference terms such as $\Omega(\Delta\hat{\varphi}/\Delta\Omega)_{T,p}$, the association of the value with a single state is somewhat arbitrary. If we label the states in the difference above as $\Omega(\Delta\hat{\varphi}/\Delta\Omega) = \Omega(\hat{\varphi}_2 - \hat{\varphi}_1)/(\Omega_2 - \Omega_1)$ we chose to assign the values to the state of $\hat{\varphi}_1, \Omega_1$. This means that $\Omega = \Omega_1$ in the expression. Another choice may be to assign the value to some mean of the states. In the limit of infinitesimal differences, both choices give the same curves. The choice we made has the advantage of being simpler to implement in software. However, if quantitative accuracy is the main concern, the second choice may be more satisfactory.

In summary we acquired simulation data for a range of pressures and areas at two different temperatures. We set the control parameters T, μ and Ω for each simulation and sampled the gas pressure and density from the region of macroscopic gas properties. The total amount of the adsorbate component was also sampled. The rest of the properties then followed by thermodynamic relations. In each simulation the system was first equilibrated for a number of cycles that depended on the state we were simulating. The properties of interest were monitored to make sure they fluctuated around a steady value. This was followed by a number of cycles where samples for calculating ensemble averages and errors were collected. The number of cycles varied with the state simulated, and were chosen in each case such that the estimated error was less than 1.5% of the mean with 95% confidence for all sampled properties and the calculated property $\bar{\Gamma}_s$. The strongest correlation between samples was found in $\bar{\Gamma}_s$. This property therefore determined the number of cycles. In estimating the standard error we accounted for correlation in the data by block analysis.

We analyzed the data using the scientific computing library SciPy [26], and created figures using Matplotlib [27]. We interpolated the simulation data for the adsorption isotherms $\bar{\Gamma}_s(p; T, \Omega)$ using univariate splines, constraining the splines to be linear for the lowest pressures according to Henry's law. We then extrapolated the linear region to zero pressure to allow the integration in Equation (46). We evaluated the integral analytically for the lowest pressures, and numerically for the remaining pressures.

5. Results

The calculated results are shown in Figures 1–4. The figures show φ, $\hat{\varphi}$ and $\mathcal{E}_s/\Omega$ and other thermodynamic properties as functions of the adsorbent area Ω at constant temperature and pressure. All quantities are given in reduced units. Figures 1 and 2 are the results for the graphite adsorbent case. Figures 3 and 4 are the results for the generic adsorbent case representing very small adsorbents with strong interaction potentials. It is clear from the figures that all properties shown depend on the system size (area). To place our experiment in context with macroscopic thermodynamics, consider an infinite flat film adsorbed on a flat adsorbent. The intensive thermodynamic properties of this film do not depend on the surface area. There is then no difference between the integral spreading pressure $\hat{\varphi}$ and the differential spreading pressure φ, and $\mathcal{E}_s = 0$. The results documented show that we are far away from this limit, since there is an observable difference between integral and differential properties. The systems can thus be considered as small in Hill's sense.

Figure 1 shows how the spreading pressures, φ and $\hat{\varphi}$, and the subdivision potential depend on system size for the graphite adsorbent case. The area range corresponds approximately to an adsorbent radius between 6 and 25 $\times 10^{-10}$ m. We can see that the trend for φ and $\hat{\varphi}$ has an initially steep increase, after which the rate of increase goes down. At the upper limit of the area range there is still a significant difference between φ and $\hat{\varphi}$. The subdivision potential per unit area $\mathcal{E}_s/\Omega$, or the deviation from the corresponding macroscopic system, appears fairly constant; however, these results are still consistent with the theoretical prediction that $\varphi = \hat{\varphi}$ and $\mathcal{E}_s$ approaches zero in the macroscopic limit; see discussion below.

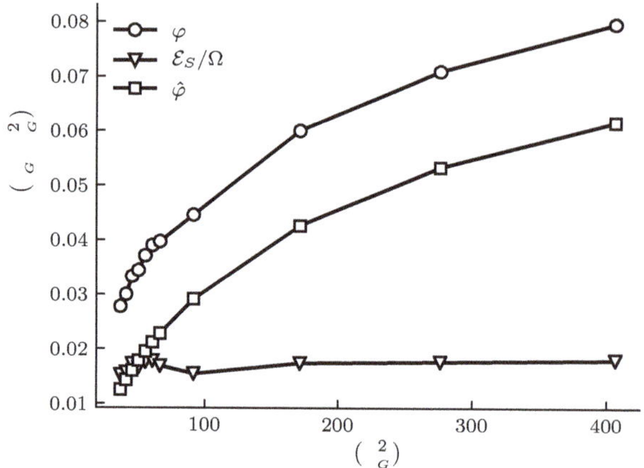

Figure 1. Film properties φ, $\hat{\varphi}$ and $\mathcal{E}_s/\Omega$ for the graphite adsorbent case as functions of the adsorbent area Ω at constant temperature $T = 1.080$ and pressure $p = 0.011$. All quantities are given in reduced units. For reference, the approximate values in SI units are $T \approx 250$ K, $p \approx 1.05$ MPa and adsorbent radii are between 6 and 25 $\times 10^{-10}$ m. The markers indicate which adsorbent sizes were simulated and the lines are there as visual aid.

Figure 2 shows the size dependence of the entropy and the enthalpy per film molecule relative to the gas, and the difference $\mathcal{E}_s/\bar{N}_s$ between the two as given by Equation (39) for the graphite adsorbent case. We see that, in general, there is less entropy and enthalpy per film molecule as the adsorbent becomes larger. The rate at which the properties decrease with increasing size goes down. The difference $\mathcal{E}_s/\bar{N}_s$, or the deviation from the corresponding macroscopic system, decreases as the adsorbent area becomes larger. This is expected.

Figure 3 shows how the spreading pressures, φ and $\hat{\varphi}$, and the subdivision potential depend on system size for the generic adsorbent case. We see that the general trend for all three curves is an initial increase to an inflection point, after which the rate of increase goes down. At the upper limit of the area range there is still a significant difference between φ and $\hat{\varphi}$, and the subdivision potential per unit area $\mathcal{E}_s/\Omega$ has not started to decrease towards zero at that point. However, the results are still consistent with the theoretical prediction that $\varphi = \hat{\varphi}$ and $\mathcal{E}_s$ approaches zero in the macroscopic limit; see discussion below. Special is a local deviation from the overall trend of the curves for adsorbent areas $50 < \Omega < 60$. This is more clearly seen in the enthalpy and entropy curves of Figure 4 for the same range of areas.

Figure 4 shows the size dependence of the entropy and the enthalpy per film molecule relative to the gas, and the difference $\mathcal{E}_s/\bar{N}_s$ between the two as given by Equation (39). We see that, in general, there is less entropy and enthalpy per film molecule as the adsorbent becomes larger. The rate at which the properties decrease with increasing size goes down. The difference $\mathcal{E}_s/\bar{N}_s$, or the deviation from the corresponding macroscopic system, decreases as the adsorbent area becomes larger. This is expected.

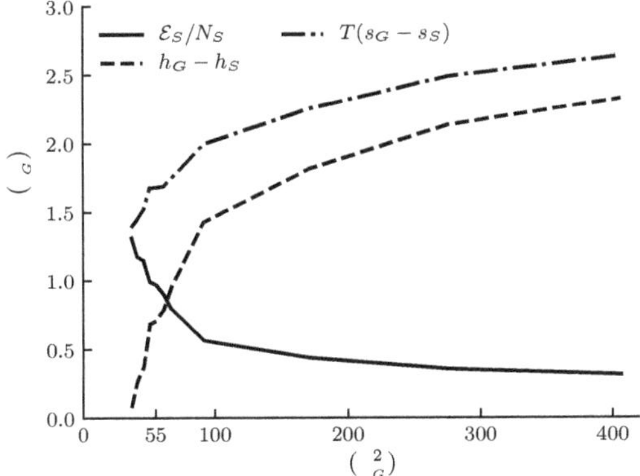

Figure 2. Film properties enthalpy, entropy and the subdivision potential per film molecule as functions of the adsorbent area Ω at constant temperature $T = 1.080$ and pressure $p = 0.011$. Enthalpy and entropy for a film molecule are given relative to the gas. All quantities are given in reduced units. For reference, the approximate values in SI units are $T \approx 250$ K, $p \approx 1.05$ MPa and adsorbent radii are between 6 and 25 $\times 10^{-10}$ m for the area range.

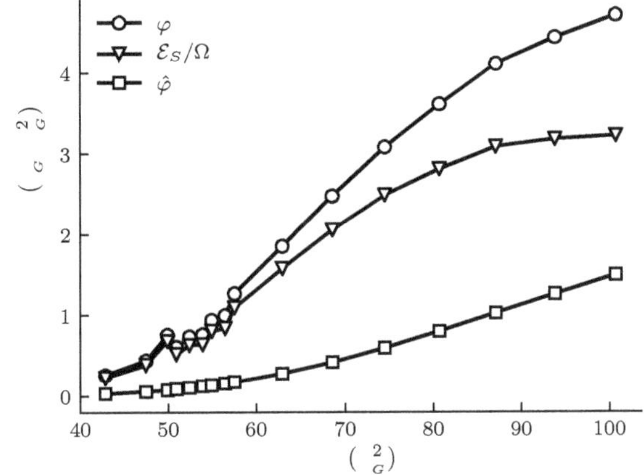

Figure 3. Film properties φ, $\hat{\varphi}$ and $\mathcal{E}_S/\Omega$ as functions of the adsorbent area Ω at constant temperature $T = 1.080$ and pressure $p = 1.8 \times 10^{-4}$ for the generic adsorbent case. All quantities are given in reduced units. For reference, the approximate values in SI units are $T \approx 250$ K, $p \approx 50$ kPa and adsorbent radii are between 6 and 9 $\times 10^{-10}$ m. The markers indicate which adsorbent sizes were simulated and the lines are there as a visual aid.

A local maximum in the entropy of the gas relative to the entropy of the film for adsorbent areas $50 < \Omega < 60$ is special, as is a corresponding maximum in the gas enthalpy relative to the film enthalpy. The quantities are plotted as $T(s_G - s_s)$ and $h_G - \bar{h}_s$ for convenience so that all quantities are positive. Thus we observed a minimum in the film entropy relative to the gas $s_s - s_G$ and a minimum in the film enthalpy relative to the gas $\bar{h}_s - h_G$. From the results, the areas around reduced units 55 appear special. Compared to Figure 2 it appears that there is local deviation from the general trend in the curves for the graphite adsorbent in this range of areas, but to a much smaller extent than for the generic adsorbent. A thermodynamic analysis alone will not reveal any particular reason for such a behavior. Several explanations can be thought of, as discussed below.

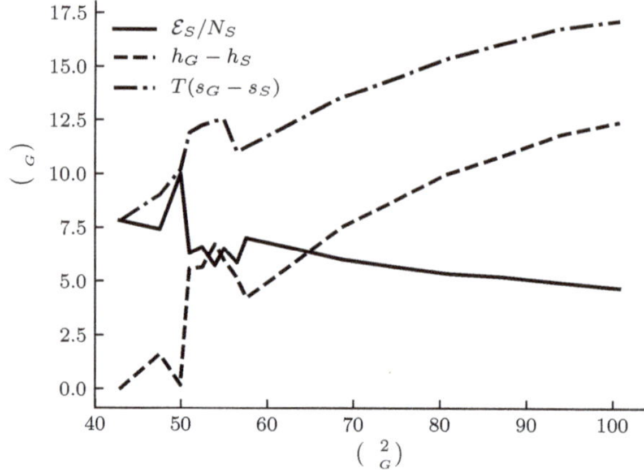

Figure 4. Film properties enthalpy, entropy and the correction function per film molecule as functions of the adsorbent area Ω at constant temperature $T = 1.080$ and pressure $p = 1.8 \times 10^{-4}$ for the generic adsorbent case. Enthalpy and entropy for a film molecule are given relative to the gas. All quantities are given in reduced units. For reference, the approximate values in SI units are $T \approx 250$ K, $p \approx 50$ kPa and adsorbent radii are between 6 and 9×10^{-10} m for the area range.

6. Discussion

In the simulations we changed the size of the adsorbent while keeping the temperature and chemical potential of the gas constant. As the gas is macroscopic, this is equivalent to controlling the temperature and pressure. In effect, as long as the gas is macroscopic it is unaffected by the size of the adsorbent. In cases where the gas is no longer macroscopic, such as in confinement, the gas phase must also be treated as a small system where all three of the variables T, p and μ are independent.

When we increase the size of the adsorbent, there are two direct changes to the composite system: (1) the strength of the external field increases, and (2) the surface becomes less curved. For the present systems, the two changes were coupled to the size of the adsorbent, so we cannot isolate the individual effects. However, if one would like to do so, one could study different adsorbent materials at the same size.

The strength of the external field felt by an adsorbed molecule comes from integrating the pairwise sum over all atoms that would constitute the adsorbent, resulting in Equation (55). Therefore, an increase in the field strength is analogous to the introduction of more adsorbent atoms. This leads to a larger total interaction energy for the film, but is to some extent counteracted by the fact that with increasing size, a larger fraction of the adsorbent atoms would be at a greater distance from a given adsorbed molecule. As the pair potential is inversely proportional to the separation distance by $u(r) \propto r^{-6}$, the greater distance leads to a plateau in the total interaction energy in the large size limit.

To further discuss the effect of the presence of the adsorbent and its size on the film, we give an illustration of four related systems divided into two groups. The first group consists of two systems that have flat surfaces, while the second group consists of two systems with curved surfaces. In each group one of the systems has the condensed phase in the presence of an adsorbent while the other has a free condensed phase. The comparison of free and adsorbed systems within a group focuses on the effects of the adsorbent, while the comparison of spherical and flat systems between the groups focuses on the effects of size. We found the analogy of the systems in the second group to the more simple systems in the first group helpful.

In the first case, consider a single-component bulk liquid with a flat surface, in equilibrium with its own vapor. The nature of this system is completely characterized by the temperature. This means that there is only one pressure $p^0(T)$ for a given temperature T at which the equilibrium state can exist.

If we now extend the definition of the system and let the liquid be adsorbed onto a flat adsorbent surface, the situation becomes different. If the pressure is below p^0 for the given temperature, the phase will start to evaporate. However, the evaporation will not necessarily continue until there is only gas left. This is because the evaporation will reach a point at which the field of force from the adsorbent is felt in such a way that no part of the phase maintains the properties of bulk. Beyond this point there is a non-zero free energy change associated with adsorption/desorption, and the amount of adsorbed material $\bar{\Gamma}_s(T, p)$ may adjust to satisfy the equilibrium condition while T and p remain independent. The condensed phase may therefore be in equilibrium with the vapor at pressures below p^0.

Now consider the analogous case of a single-component spherical liquid droplet of area Ω^* in equilibrium with its own vapor. The (unstable) equilibrium state is characterized by T and Ω^*. This means that for a given temperature and area there is only one pressure $p^{0'} > p^0$ at which the equilibrium state can exist. The equilibrium pressure $p^{0'}(T, \Omega^*)$ for this system is larger than p^0 due to the curvature and the Laplace pressure. If we let the liquid be an adsorbed phase on a spherical adsorbent, as in our simulations, the situation changes again. The system has three independent variables. The equilibrium state may be characterized by T, Ω and p as in Equation (21). Compared to the droplet at the same temperature, we set the adsorption such that the two physical systems are equal in size as characterized by Ω^*. There is one more independent variable to fix. Therefore, unlike for the droplet we have many possible equilibrium pressures for a given adsorption (as determined by Ω^*) if we let Ω be different in each case. In Figures 1–4 we have many adsorptions for a given pressure because the area is different in each case. For given values of T and Ω there is a limit pressure $p^{0'}$ at which bulk condensation starts. This limit becomes lower for larger Ω until it reaches p^0 for a large flat adsorbent. In other words, a given adsorbent size imposes a limit on the vapor pressure required for bulk condensation. This has previously been observed with a statistical model [28]. As the limit changes with size, this implies that a change in size of the adsorbent for $p^0 < p < p^{0'}$ may induce bulk condensation if $p^{0'}$ drops below p. For $p < p^0$ there is naturally no adsorbent size that will induce bulk condensation. After bulk condensation starts, the saturation pressure decreases gradually towards p^0 because the curvature of the growing condensed phase decreases and approaches zero.

To summarize, for the first system we only have to fix the temperature in order to fix the equilibrium state. In the second system we have, in addition, to fix the pressure. In the third system we have to fix the temperature and the area of the adsorbent. In the last system we have to fix the temperature, the area and the pressure.

For our simulations the illustration above implies that for pressures $p < p^0$, we may have a flat and a spherical adsorbed film in equilibrium at the same temperature and pressure. They will in general have different thermodynamic properties. For instance, the adsorption differs as discussed above. By increasing the adsorbent size, the bulk condensation pressure limit approaches p^0 from above and the thermodynamic properties approach the values for a flat film. For pressures $p > p^0$, the spherical film may be in equilibrium but there is no possible corresponding flat system because $p^0 < p^{0'}$. An increase of adsorbent size may cause the bulk condensation pressure limit $p^{0'}$ to pass through p, at which point the adsorption diverges.

In Figures 1 and 3 the pressures are below p^0 for the given temperature. We would therefore expect the curves for φ and $\hat{\varphi}$ to eventually coincide at the plateau value for a flat adsorbed film at the given temperature and pressure. This occurs when the slope of the $\hat{\varphi}$ curve is zero, which is consistent with the relation $\varphi = \hat{\varphi} + \Omega(\partial\hat{\varphi}/\partial\Omega)_{T,p}$ and Equation (44) as $\mathcal{E}_s = 0$. The curve for $\hat{\varphi}$ in Figure 3 is s-shaped and consistent with the approach to a plateau value. According to Equation (44), the s-shaped curve of $\hat{\varphi}$ gives a bell-shaped curve for $\mathcal{E}_s/\Omega$, with a peak value at the inflection point of $\hat{\varphi}$, and ending at $\mathcal{E}/\Omega = 0$. The expected curve for φ would according to the relation $\varphi = \hat{\varphi} + \Omega(\partial\hat{\varphi}/\partial\Omega)_{T,p}$ also have a peak before decreasing towards the plateau value of $\hat{\varphi}$. The same reasoning applies to Figure 1, however the shapes of the curves are stretched out and harder to identify. Thus, although the eventual decrease in $\mathcal{E}_s$ towards zero is not observed for the limited area ranges in Figures 1 and 3, the results are consistent with the macroscopic limit relations $\varphi = \hat{\varphi}$ and $\mathcal{E}_s = 0$. More experiments are needed for further confirmations.

The trends in the entropy and enthalpy curves in Figures 2 and 4 are reasonable, when we consider the discussion above on the change in total interaction energy with size. With a larger interaction energy, the molecules are more tightly bound to the surface, and there are less possible configurations of the system available. The adsorbed phase is then more ordered, and the entropy becomes lower. In the macroscopic limit where the film is flat $\mathcal{E}_s / \bar{N}_s = 0$ and $\bar{h}_s - h_G = T(s_s - s_G)$ by Equation (39). The trends of the curves for $\mathcal{E}_s / \bar{N}_s$ are consistent with this. More experiments are needed for further confirmations.

For the very small adsorbent with a strong interaction potential(cf. Figure 4), we can observe a peak in the entropy and enthalpy, and there is a corresponding effect in Figure 3. A minimum in the entropy of an adsorbed molecule, when plotted against the adsorption or pressure, is well known in systems where the interaction of the adsorbate with the adsorbent is strong. This is related to the decrease in the number of configurations available to the system as the first adsorption layer is filled and a subsequent increase following the initiation of adsorption in a second layer. No film density variation consistent with this was observed. The peak we see in Figure 4 was also observed for the same areas for a range of different adsorptions, indicating that it does not have the same origin as discussed above. These were observations from curves showing the same qualitative behavior but at different pressures than the ones shown here. The start of the second layer was estimated by the location of the minimum in the entropy when plotted against the adsorption (not shown here). The local minimum we observe may thus best be related to available configurations of the system in other ways, such as the change in structure and arrangement of the molecules on the surface. The type of entropy we are showing in Figure 4 is the entropy per molecule $S_s / \bar{N}_s$ (relative to the gas). As emphasized by Hill [16], this type of entropy, as opposed to the differential entropy $(\partial S_s / \partial \bar{N}_s)_{T,\Omega}$ is the correct one to discuss in relation to the degree of order of the adsorbed molecules. More information on the spatial arrangement of the adsorbed molecules may allow this question to be investigated further.

7. Conclusions and Perspectives

We have seen in this work how we can deal with size as a variable in a systematic manner. The common thermodynamic equations for adsorption were changed in three ways.

Using the procedure of Hill, we have calculated the integral spreading pressure from adsorption isotherm data for a fixed adsorbent size, using Equation (46). The equation has the same form as the one usually encountered in adsorption thermodynamics, the integral form of Equation (26) in [13], with the important distinction that $\hat{\varphi}$ and φ in Hill's description are not the same functions for small systems. Equation (46) is valid for systems of any size.

Similarly, we have shown how to obtain the entropy and the enthalpy per film molecule by Equations (36) and (39). Equation (36) is the same as the one usually encountered in adsorption thermodynamics, Equation (21) in [13], except for the important fact that the function that is kept constant when we take the derivative is now $\hat{\varphi}$ instead of φ.

This entropy is the entropy typically discussed in relation to the degree of order of the adsorbed molecules. Equation (39) differs from the usual equation by the term $\mathcal{E}_s / \bar{N}_s$. We observe that for a small system $\bar{h}_s - h_G \neq T(s_s - s_G)$ because $\mathcal{E}_s \neq 0$. The expressions referred to above, become the usual expressions in the macroscopic limit where $\mathcal{E}_s = 0$ and $\hat{\varphi} = \varphi$.

The effects of size on intensive variables is a feature of Hill's nanothermodynamics [17], and may be expressed in terms of $\mathcal{E}_s$ and derivatives of $\mathcal{E}_s$ by Equations (41)–(45). By the subdivision potential, we have a direct measure of the system smallness, and we can see from the numerical data that the value is significant in all relevant properties.

On this basis, we can suggest some possible directions for future work. The adsorbent may be modeled explicitly as a collection of atoms instead of as an integrated potential. This may allow the investigation of adsorption on a non-smooth surface with adsorption sites. By allowing the adsorbent to be compressible, it may be possible to observe how the film is affected by a phase transition in the adsorbent for different sizes. By modeling the adsorbate as all-atom molecules instead of coarse-grained particles, it may be possible to study changes in the molecular orientation and structure of adsorbing layers induced by the adsorbent size.

Author Contributions: All work not specified was done by B.A.S. Supervision, review, project administration and funding acquisition was done by J.H. Critical review and editing was contributed by D.B. and S.K. All authors contributed to the scientific discussion. All authors have read and agreed to the published version of the manuscript.

Funding: B.A.S. and J.H. acknowledge funding by the Research Council of Norway, Det Norske Oljeselskap ASA, Wintershall Norge AS via WINPA project (Nano2021 and Petromaks2 234626), and the Norwegian Metacenter for Computational Science (grant numbers NN9110k and NN9391k). D.B. and S.K. acknowledge funding by the Research Council of Norway through its Center of Excellence Funding Scheme, project number 262644, PoreLab.

Acknowledgments: We are grateful to Jean-Marc Simon for commenting on the manuscript.

Conflicts of Interest: The authors declare no conflict of interest.

References

1. Boucher, O.; Randall, D.; Artaxo, P.; Bretherton, C.; Feingold, G.; Forster, P.; Kerminen, V.; Kondo, Y.; Liao, H.; Lohmann, U.; et al. Clouds and Aerosols. In *Climate Change 2013: The Physical Science Basis. Contribution of Working Group I to the Fifth Assessment Report of the Intergovernmental Panel on Climate Change*; Cambridge University Press: Cambridge, UK, 2013; pp. 571–657.
2. Tolman, R.C. The Effect of Droplet Size on Surface Tension. *J. Chem. Phys.* **1949**, *17*, 333–337. [CrossRef]
3. Rusanov, A.I. The thermodynamics of processes of new-phase formation. *Russ. Chem. Rev.* **1964**, *33*, 385–399. [CrossRef]
4. Schmelzer, J.W.; Gutzow, I.; Schmelzer, J., Jr. Curvature-Dependent Surface Tension and Nucleation Theory. *J. Colloid Interface Sci.* **1996**, *178*, 657–665. [CrossRef]
5. Reguera, D.; Reiss, H. Nucleation in confined ideal binary mixtures: The Renninger–Wilemski problem revisited. *J. Chem. Phys.* **2003**, *119*, 1533–1546. [CrossRef]
6. Schmelzer, J.W.P.; Boltachev, G.S.; Baidakov, V.G. Classical and generalized Gibbs' approaches and the work of critical cluster formation in nucleation theory. *J. Chem. Phys.* **2006**, *124*, 194503. [CrossRef] [PubMed]
7. Dubrovskii, V.G. Refinement of Nucleation Theory for Vapor–Liquid–Solid Nanowires. *Cryst. Growth Des.* **2017**, *17*, 2589–2593. [CrossRef]
8. Aasen, A.; Blokhuis, E.M.; Wilhelmsen, Ø. Tolman lengths and rigidity constants of multicomponent fluids: Fundamental theory and numerical examples. *J. Chem. Phys.* **2018**, *148*, 204702. [CrossRef] [PubMed]
9. Gibbs, J.W. *The Scientific Papers of J. Willard Gibbs, Volume 1, Thermodynamics*; Ox Bow Press: Woodbridge, CT, USA, 1993.
10. Hill, T.L. Thermodynamics of Small Systems. *J. Chem. Phys.* **1962**, *36*, 3182. [CrossRef]
11. Bedeaux, D.; Kjelstrup, S. Hill's nano-thermodynamics is equivalent with Gibbs' thermodynamics for surfaces of constant curvatures. *Chem. Phys. Lett.* **2018**, *707*, 40–43. [CrossRef]
12. Fowler, R.H.; Guggenheim, E.A. *Statistical Thermodynamics: A Version of Statistical Mechanics for Students of Physics and Chemistry*; Macmillan: New York, NY, USA, 1939.
13. Hill, T.L. Statistical Mechanics of Adsorption. IX. Adsorption Thermodynamics and Solution Thermodynamics. *J. Chem. Phys.* **1950**, *18*, 246–256. [CrossRef]
14. Guggenheim, E.A. *Modern Thermodynamics by the Methods of Willard Gibbs*; Methuen & Company Limited: London, UK, 1933; p. 166.
15. Bedeaux, D.; Kjelstrup, S.; Schnell, S.K. *Nanothermodynamics—General Theory*; lPoreLab: Trondheim, Norway, 2020.
16. Hill, T.L. Statistical Mechanics of Adsorption. V. Thermodynamics and Heat of Adsorption. *J. Chem. Phys.* **1949**, *17*, 520–535. [CrossRef]
17. Hill, T.L. *Thermodynamics of Small Systems*; Dover Books on Chemistry; Dover Publications: New York, NY, USA, 1963.
18. Hill, T. *An Introduction to Statistical Thermodynamics*; Dover Books on Physics; Dover Publications: New York, NY, USA, 2012.

19. Strøm, B.A.; Simon, J.M.; Schnell, S.K.; Kjelstrup, S.; He, J.; Bedeaux, D. Size and shape effects on the thermodynamic properties of nanoscale volumes of water. *Phys. Chem. Chem. Phys.* **2017**, *19*, 9016–9027. [CrossRef] [PubMed]
20. Rauter, M.T.; Galteland, O.; Erdős, M.; Moultos, O.A.; Vlugt, T.J.H.; Schnell, S.K.; Bedeaux, D.; Kjelstrup, S. Two-Phase Equilibrium Conditions in Nanopores. *Nanomaterials* **2020**, *10*, 608. [CrossRef] [PubMed]
21. Myers, A.L.; Prausnitz, J.M. Thermodynamics of mixed-gas adsorption. *AIChE J.* **1965**, *11*, 121–127. [CrossRef]
22. Plimpton, S. Fast Parallel Algorithms for Short-Range Molecular Dynamics. *J. Comput. Phys.* **1995**, *117*, 1–19. [CrossRef]
23. Mie, G. Zur kinetischen Theorie der einatomigen Körper. *Ann. Phys.* **1903**, *316*, 657–697. [CrossRef]
24. Avendaño, C.; Lafitte, T.; Galindo, A.; Adjiman, C.S.; Jackson, G.; Müller, E.A. SAFT-γ Force Field for the Simulation of Molecular Fluids. 1. A Single-Site Coarse Grained Model of Carbon Dioxide. *J. Phys. Chem. B* **2011**, *115*, 11154–11169. [CrossRef]
25. Graham, S.C.; Homer, J.B.; Rosenfeld, J.L.J. The Formation and Coagulation of Soot Aerosols Generated by the Pyrolysis of Aromatic Hydrocarbons. *Proc. R. Soc. London. Ser. A Math. Phys. Sci.* **1975**, *344*, 259–285.
26. Vijaykumar, A.; Bardelli, A.P.; Rothberg, A.; Hilboll, A.; Kloeckner, A.; Scopatz, A.; Lee, A.; Rokem, A.; Woods, N.; Fultonet, C.; et al. SciPy 1.0: Fundamental Algorithms for Scientific Computing in Python. *Nat. Methods* **2020**, *17*, 261–272. [CrossRef]
27. Hunter, J.D. Matplotlib: A 2D Graphics Environment. *Comput. Sci. Eng.* **2007**, *9*, 90–95. [CrossRef]
28. Hill, T.L. The Note on the Physical Adsorption of Gases in Capillaries and on Small Particles (Nucleation of Condensation). *J. Phys. Colloid Chem.* **1950**, *54*, 1186–1191. [CrossRef] [PubMed]

 *nanomaterials*

MDPI

Article

Adsorption of an Ideal Gas on a Small Spherical Adsorbent

Bjørn A. Strøm [1,*], **Dick Bedeaux** [2] **and Sondre K. Schnell** [1]

[1] Department of Materials Science and Engineering, Faculty of Natural Sciences, Norwegian University of Science and Technology, NTNU, NO-7491 Trondheim, Norway; sondre.k.schnell@ntnu.no

[2] Porelab, Department of Chemistry, Norwegian University of Science and Technology, NTNU, NO-7491 Trondheim, Norway; dick.bedeaux@ntnu.no

[*] Correspondence: bjorn.a.strom@ntnu.no

Abstract: The ideal gas model is an important and useful model in classical thermodynamics. This remains so for small systems. Molecules in a gas can be adsorbed on the surface of a sphere. Both the free gas molecules and the adsorbed molecules may be modeled as ideal for low densities. The adsorption energy, U^s, plays an important role in the analysis. For small adsorbents this energy depends on the curvature of the adsorbent. We model the adsorbent as a sphere with surface area $\Omega = 4\pi R^2$, where R is the radius of the sphere. We calculate the partition function for a grand canonical ensemble of two-dimensional adsorbed phases. When connected with the nanothermodynamic framework this gives us the relevant thermodynamic variables for the adsorbed phase controlled by the temperature T, surface area Ω, and chemical potential μ. The dependence of intensive variables on size may then be systematically investigated starting from the simplest model, namely the ideal adsorbed phase. This dependence is a characteristic feature of small systems which is naturally expressed by the subdivision potential of nanothermodynamics. For surface problems, the nanothermodynamic approach is different, but equivalent to Gibbs' surface thermodynamics. It is however a general approach to the thermodynamics of small systems, and may therefore be applied to systems that do not have well defined surfaces. It is therefore desirable and useful to improve our basic understanding of nanothermodynamics.

Keywords: adsorption; nanothermodynamics; small-system; size-dependent; thermodynamics; statistical mechanics; ideal gas; nanoparticles

 check for
updates

Citation: Strøm, B.A.; Bedeaux, D.; Schnell, S.K. Adsorption of an Ideal Gas on a Small Spherical Adsorbent. *Nanomaterials* **2021**, *11*, 431. https:// doi.org/10.3390/nano11020431

Academic Editor: Abdelhamid Elaissari

Received: 21 January 2021
Accepted: 6 February 2021
Published: 9 February 2021

Publisher's Note: MDPI stays neutral with regard to jurisdictional claims in published maps and institutional affiliations.

1. Introduction

The objective of the paper is to demonstrate an organized and transparent thermodynamic framework for statistical model development for small systems. The main focus from the thermodynamic side is on the characteristic feature of small systems, namely the effect of size on intensive variables.

We do this by first obtaining the characteristic thermodynamic function for the adsorbed phase from nanothermodynamics as introduced by Hill [1–3]. This function is the one that provides us with the fundamental link to statistical mechanics, as it is equal to $-kT$ times the logarithm of the grand canonical partition function. The characteristic function for the adsorbed phase depends on the size Ω of the system. In the macroscopic limit the dependence becomes linear, however when the system is small the subdivision potential measures the deviation from macroscopic behavior. We therefore derive an expression for the subdivision potential in terms of the environment variables, and observe that the differential coefficients of this expression give the dependence of particular intensive properties on size. The size dependence of other intensive thermodynamic properties may then be expressed through thermodynamic relations in terms of the subdivision potential and its derivatives. The close relationship between the subdivision potential and the characteristic feature of small systems is a consequence of the generalization of thermodynamics to small systems, and the framework's internal structure that follows. This is what we wish to emphasize in this work. For surface problems, the nanothermodynamic approach is different, but equivalent to Gibbs' surface thermodynamics [1,4,5].

Since the quantities usually referred to as intensive now depend on size, the classical meaning of the term intensive is not appropriate for small systems. It is still a useful term to distinguish thermodynamic quantities, especially since nanothermodynamics is a generalization of classical thermodynamics, and therefore goes over into classical thermodynamics in the macroscopic limit, where this term is ingrained. For a small system quantity, the term intensive is used to describe a quantity that becomes intensive in the classical sense in the macroscopic limit.

We give the thermodynamic framework substance by calculating the subdivision potential of the adsorbed phase. Taking advantage of the simplicity of the ideal gas model, the thermodynamic quantities become more tangible, and it is possible later to gradually increase the complexity from this model to include effects like crowding and cooperativity.

The model consists of an adsorbed phase that is ideal and has an adsorption energy U^s. The adsorption energy U^s will in general depend on the curvature of the sphere, the temperature T and the chemical potential μ. However, we want the simplest model possible, in order to make the connection between the nanothermodynamic framework and statistical mechanics as clear and minimal as possible. We therefore consider U^s to only depend on Ω, which is consistent with an inert and incompressible adsorbent. The control variables are therefore T, Ω, and μ. The dependence on the curvature is characteristic for small spheres.

If the structure of the adsorbent is taken into account, and different crystal structures are considered, this will result in different size dependence for the intensive variables of the adsorbed phase. This is because the surface to volume ratio of the adsorbent becomes a different function of size, and also because edges and corners will have to be considered. This is interesting, but comes at the expense of increased complexity, and is beyond the scope of the article.

2. Nanothermodynamics

The thermodynamic system considered here is the adsorbed phase in the context of adsorption of a single component gas on an inert adsorbent. The adsorbent is assumed to be unaffected by the temperature, chemical potential and the adsorbed layer. It functions only as an external field, and is therefore not included in the description of the system. The adsorbed phase is in equilibrium with the gas. The temperature T, chemical potential μ, and surface area of the sphere $\Omega = 4\pi R^2$ form a complete set of independent variables for the adsorbed phase. The surface area Ω determines the radius $R = \sqrt{\Omega/4\pi}$ and the curvature $C = 2/R$. The curvature dependence of the surface tension can therefore be written as a dependence on the surface area Ω. All thermodynamic quantities of the adsorbed phase are functions of T, μ, and Ω.

Following Hill [2,3] we consider an ensemble of $\mathcal{N}$ independent small systems at temperature T, and component chemical potential μ. A complete set of independent variables for the ensemble, with total properties denoted by subscript t, may then be taken as the entropy S_t, area $\mathcal{N}\Omega$, the amount of adsorbed component N_t, and the number or replicas $\mathcal{N}$. We note here that we allow for an independent variation in the size of the small systems, as given by Ω, in addition to the variation in the number of small systems $\mathcal{N}$. This is an essential new feature that allows us to investigate the size of the small system the ensemble represents, and which makes the approach distinct from simply describing a large sample of small systems by conventional thermodynamics.

The characteristic function for the ensemble in terms of the set of independent variables S_t, Ω, N_t, and $\mathcal{N}$ is the internal energy U_t given by

$$dU_t = TdS_t + \gamma\mathcal{N}d\Omega + \mu dN_t + Xd\mathcal{N} \tag{1}$$

where γ is the ensemble mean surface tension. We will refer to this equation as the Hill-Gibbs equation. The intensive quantities are given by

$$T = \left(\frac{\partial U_t}{\partial S_t}\right)_{\Omega,N_t,\mathcal{N}}, \quad \gamma\mathcal{N} = \left(\frac{\partial U_t}{\partial \Omega}\right)_{S_t,N_t,\mathcal{N}}, \quad \mu = \left(\frac{\partial U_t}{\partial N_t}\right)_{S_t,\Omega,\mathcal{N}} \tag{2}$$

And finally we have the so-called replica energy

$$X = \left(\frac{\partial U_t}{\partial \mathcal{N}}\right)_{S_t,\Omega,N_t} \tag{3}$$

This energy is needed when one ads a replica with a surface area Ω while redistributing the total adsorbed entropy and number of particles over one more replica. Using that U_t, S_t, N_t are Euler homogeneous in the number of replica $\mathcal{N}$ ([2]), i.e., proportional to $\mathcal{N}$, it follows that

$$U_t = TS_t + \mu N_t + X\mathcal{N} \tag{4}$$

The internal energy, entropy and number of particles of the adsorbed phase per replica are defined by

$$U \equiv U_t/\mathcal{N}, \quad S \equiv S_t/\mathcal{N}, \quad N \equiv N_t/\mathcal{N} \tag{5}$$

Apart from the entropy S the quantities defined in Equation (5) are ensemble mean values of fluctuating extensive quantities ([2], p. 9), ([6], p. 98). For small systems, if a quantity does not fluctuate, but has the same value in every system of the ensemble, it is an environment variable. Here these variables are T, Ω and μ. The exception to this rule is the entropy, which is a property of the complete distribution in internal energy and particle number for a single system ([2], p. 9), and is therefore the same for each system. Together with Equation (4) it follows that

$$\hat{\gamma}\Omega \equiv X = U - TS - \mu N \tag{6}$$

This equation also defines $\hat{\gamma}$, the characteristic energy per unit area. Substitution of Equation (5) into Equation (1) and using Equation (6) gives the Gibbs equation for the replicas

$$dU = TdS + \gamma d\Omega + \mu dN \tag{7}$$

The important difference of this equation with the usual Gibbs equation is that U, S, N are, for a small sphere, not Euler homogeneous in the surface area Ω, i.e., not proportional to Ω. Differentiating Equation (6) and using Equation (7) we obtain what we call the Hill-Gibbs-Duhem equation

$$d(\hat{\gamma}\Omega) = -SdT + \gamma d\Omega - Nd\mu \tag{8}$$

It follows from this equation that

$$\left(\frac{\partial \hat{\gamma}}{\partial T}\right)_{\Omega,\mu} = -\frac{S}{\Omega} \equiv -s, \quad \left(\frac{\partial \hat{\gamma}}{\partial \mu}\right)_{T,\Omega} = -\frac{N}{\Omega} \equiv -n$$
$$\left(\frac{\partial(\hat{\gamma}\Omega)}{\partial \Omega}\right)_{T,\mu} = \hat{\gamma} + \Omega\left(\frac{\partial \hat{\gamma}}{\partial \Omega}\right)_{T,\mu} = \gamma \tag{9}$$

We define the subdivision potential $\mathcal{E}$ by

$$\mathcal{E} \equiv (\hat{\gamma} - \gamma)\Omega \tag{10}$$

While the form and physical significance of Equation (7) is the same for small and large systems, we see by using Equations (6) and (10) that the Euler equation for a small system takes a different form

$$U = TS + \gamma\Omega + \mu N + \mathcal{E} \tag{11}$$

which shows the central role of the subdivision potential. Equation (7) together with Equation (8) gives

$$d\mathcal{E} = -SdT - \Omega d\gamma - Nd\mu \tag{12}$$

Using the Gibbs-Duhem equation in the large surface area (thermodynamic) limit it follows that $\mathcal{E} = 0$ in this limit. This implies that $\widehat{\gamma} = \gamma$ in the thermodynamic limit. Equation (12) shows furthermore that

$$\left(\frac{\partial\mathcal{E}}{\partial T}\right)_{\gamma,\mu} = -S, \quad \left(\frac{\partial\mathcal{E}}{\partial\gamma}\right)_{T,\mu} = \Omega, \quad \left(\frac{\partial\mathcal{E}}{\partial\mu}\right)_{T,\gamma} = -N \tag{13}$$

The intensive variables T, γ, μ determine all the extensive variables S, Ω, N. This is possible for a small sphere, and is a feature specific to small systems. In the large sphere limit S, Ω, N all become infinitely large. The change in the subdivision potential may be written in a form more appropriate for the environment the small system is in

$$d\mathcal{E} = \Omega^2\left(\frac{\partial s}{\partial\Omega}\right)_{T,\mu} dT - \Omega\left(\frac{\partial\gamma}{\partial\Omega}\right)_{T,\mu} d\Omega + \Omega^2\left(\frac{\partial n}{\partial\Omega}\right)_{T,\mu} d\mu \tag{14}$$

The effects of size on intensive variables, a characteristic feature of small systems, are now directly available as the differential coefficients of Equation (14). This relation is especially useful because the independent variables are the environment variables.

3. The Model

The one-particle canonical partition function for a small sphere with surface adsorption follows from statistical mechanics [7]:

$$Q_1(T,\Omega) = \frac{\Omega}{\Lambda^2}\exp(-\beta U^s) \tag{15}$$

where U^s is the potential energy of interaction between the adsorbent and an adsorbed molecule, $\Lambda \equiv \sqrt{h^2/(2\pi m k_\mathrm{B}T)}$ is the mean thermal de Broglie wave length. Here m is the particle mass. The N-particle canonical partition function becomes:

$$Q(T,\Omega,N) = \frac{1}{N!}Q_1^N(T,\Omega) \tag{16}$$

The grand canonical partition function equals

$$\begin{aligned}
\Xi(T,\Omega,\mu) &= \sum_{N=0}^{\infty}\exp(\beta\mu N)Q(T,\Omega,N) \\
&= \exp(\exp(\beta\mu)Q_1(T,\Omega)) \\
&= \exp\left\{\frac{\Omega}{\Lambda^2}\exp[\beta(\mu - U^s)]\right\}
\end{aligned} \tag{17}$$

where we used Equation (16). By introducing the expressions above, thermodynamic properties can be derived in terms of T, Ω, μ. From Equation (17) we find for the integral surface tension

$$\widehat{\gamma} = -\frac{k_\mathrm{B}T}{\Omega}\ln\Xi(T,\Omega,\mu) = -\frac{k_\mathrm{B}T}{\Lambda^2}\exp[\beta(\mu - U^s)] \tag{18}$$

This is the equation of state for the adsorbed phase controlled by the grand canonical ensemble. The differential surface tension is given by Equation (9)

$$\gamma = -\frac{k_B T}{\Lambda^2} \exp[\beta(\mu - U^s)]\left(1 - \beta\Omega\left(\frac{\partial U^s}{\partial \Omega}\right)_{T,\mu}\right) \tag{19}$$

By using Equation (10), we can now determine the subdivision potential:

$$\mathcal{E} = -\left(\frac{\Omega}{\Lambda^2}\right)\exp[\beta(\mu - U^s)]\Omega\left(\frac{\partial U^s}{\partial \Omega}\right)_{T,\mu} \tag{20}$$

In the thermodynamic limit U^s becomes independent of Ω so that $\mathcal{E}$ approaches zero. The entropy density $s = S/\Omega$ becomes using Equation (9)

$$s = (k_B/\Lambda^2)\exp[\beta(\mu - U^s)][2 - \beta(\mu - U^s)] \tag{21}$$

The particle density $n = N/\Omega$ becomes using Equation (9)

$$n = (1/\Lambda^2)\exp[\beta(\mu - U^s)] \tag{22}$$

Thermodynamic quantities of the adsorbed phase may be expressed per molecule. The quantities are then given by particularly simple expressions. It follows from Equations (18)–(22) that

$$\frac{\hat{\gamma}\Omega}{N} = -k_B T \tag{23}$$

$$\frac{\gamma\Omega}{N} = -k_B T + \Omega\left(\frac{\partial U^s}{\partial \Omega}\right)_{T,\mu} \tag{24}$$

$$\frac{\mathcal{E}}{N} = -\Omega\left(\frac{\partial U^s}{\partial \Omega}\right)_{T,\mu} \tag{25}$$

$$\frac{S}{N} = k_B[2 - \beta(\mu - U^s)] \tag{26}$$

$$\frac{U}{N} = \frac{X}{N} + T\frac{S}{N} + \mu = -k_B T + k_B T[2 - \beta(\mu - U^s)] + \mu = k_B T + U^s \tag{27}$$

For the differential entropy and internal energy we have the model expressions

$$\left(\frac{\partial S}{\partial N}\right)_{T,\Omega} = \left(\frac{\partial s}{\partial \mu}\right)_{T,\Omega}\bigg/\left(\frac{\partial n}{\partial \mu}\right)_{T,\Omega} \tag{28}$$

$$= k_B[1 - \beta(\mu - U^s)]$$

$$\left(\frac{\partial U}{\partial N}\right)_{T,\Omega} = k_B T + U^s \tag{29}$$

4. Correspondence with Experiment

Although comparisons with experimental results are not part of this work, the reader may be interested in the relevant relations. Furthermore, the connection to experiment may help make the description less abstract, so we allow ourselves this small detour here. We are mainly interested in how the thermodynamic properties of the adsorbed phase are affected when we vary Ω. The experimental system is typically a large collection of spherical adsorbents, such as a powder, in equilibrium with the adsorbate gas. From an experimental perspective, by using the environment variable Ω we imply that the small system is rigid. This is because we do not have any direct means of controlling the adsorbent size. We control the system experimentally through the surrounding gas. Thus, if the adsorbent is not rigid, we cannot prevent the adsorbent size from fluctuating. Instead we treat Ω as a variable parameter, and we control Ω by performing experimental measurements on (monodisperse) samples prepared with different values of Ω.

In order to assess the statistical model we may derive relations connecting thermodynamic properties of the adsorbed phase to experimentally convenient variables, see [8,9], and Appendix A. We may asses the adsorbed phase entropy and energy per molecule, and the differential entropy and energy (all relative to the gas) by

$$\frac{S}{N} - s^G = -kT\left(\frac{\partial \ln p}{\partial T}\right)_{\hat{\gamma},\Omega} \tag{30}$$

$$\frac{U}{N} - u^G = kT\left[1 - T\left(\frac{\partial \log p}{\partial T}\right)_{\hat{\gamma},\Omega}\right] \tag{31}$$

$$\left(\frac{\partial S}{\partial N}\right)_{T,\Omega} - s^G = -kT\left(\frac{\partial \ln p}{\partial T}\right)_{\Omega,N} \tag{32}$$

$$\left(\frac{\partial U}{\partial N}\right)_{T,\Omega} - u^G = kT\left[1 - T\left(\frac{\partial \log p}{\partial T}\right)_{\Omega,N}\right] \tag{33}$$

where $s^G \equiv S^G/N^G$ is the gas entropy S^G per gas molecule N^G, and $u^G \equiv U^G/N^G$ is the gas internal energy U^G per gas molecule. These are only a selection of relations that may be useful.

5. The Potential U^s

The adsorbent functioning as an external field was represented by a sphere of uniform density ρ and radius a, see Figure 1. The total interaction energy $\mathcal{U}$ was determined by integrating the interaction energy $4\pi\rho u(r_{LJ})a'^2\,da'$ between a volume element of the adsorbent, and a gas molecule separated by the distance r_{LJ}. The interaction potential $u(r_{LJ})$ was given by the standard Lennard-Jones 12-6 potential:

$$u(r_{LJ}) = 4\epsilon\left[\left(\frac{\sigma}{r_{LJ}}\right)^{12} - \left(\frac{\sigma}{r_{LJ}}\right)^{6}\right] \tag{34}$$

where ϵ is the energy parameter of the interaction, and σ is the length parameter of the interaction. We used reduced units; ϵ as the unit of energy, σ as the unit of length, the gas molecular mass as the unit of mass, and $k_B = 1$.

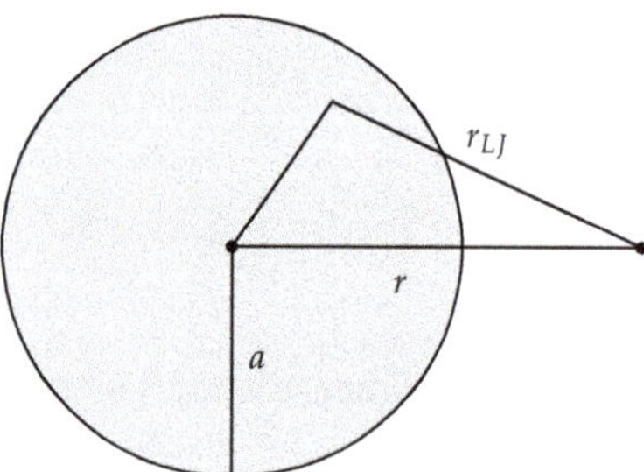

Figure 1. Illustration of the adsorbent with radius a. The distance between a volume element of the adsorbent and a gas molecule is given by r_{LJ}, and the distance between the adsorbent center and the same gas molecule is given by r.

Integrating Equation (34) over the spherical adsorbent we have

$$\mathcal{U}(a,r) = \frac{16\pi\epsilon\rho\sigma^3}{3}\left[\frac{(15a^3 r^6 + 63a^5 r^4 + 45a^7 r^2 + 5a^9)\sigma^9}{15(r^2 - a^2)^9} - \frac{a^3\sigma^3}{(r^2 - a^2)^3}\right], \quad r > a \tag{35}$$

where r is the center to center distance between the adsorbent and a gas molecule. The location of an adsorbed molecule relative to the center of the adsorbent is R, which is determined by the control variable Ω by the equation $\Omega = 4\pi R^2$. Operationally it is more practical to control a and determine Ω by a dividing surface condition involving a, than it is the other way around. The correspondence between R and a is established by the condition $\mathcal{U}(a, r = R) = \min[\mathcal{U}(a, r)]$, i.e., for a given adsorbent size a, the location R of the dividing surface is the location of the minimum of the potential $\mathcal{U}$. Thus, the adsorption energy U^s of an adsorbed molecule is determined by a, Equation (35), the fixed chosen condition, and a fixed value of ρ. We choose $\rho = 1 / \left(\left[4\pi(\sigma/2)^3 \right] / 3 \right)$.

6. Results

In this section we present calculations for the ideal adsorbed phase to show the size dependence of some important intensive properties, and give some substance to the thermodynamic framework.

Figure 2 shows the integral surface tension $\widehat{\gamma}$ which is the characteristic energy per unit area, the differential surface tension γ, and the subdivision potential per unit area $\mathcal{E}/\Omega$, as functions of the adsorbent radius a at constant temperature and chemical potential. The names integral and differential are here used to refer to the relation between $\widehat{\gamma}$ and γ in Equation (9). The quantities $\widehat{\gamma}$, γ and $\mathcal{E}$ are calculated by Equations (18)–(20). We observe that when the system becomes larger, $\mathcal{E}$ approaches zero, and $\widehat{\gamma}$ and γ both approach a plateau value.

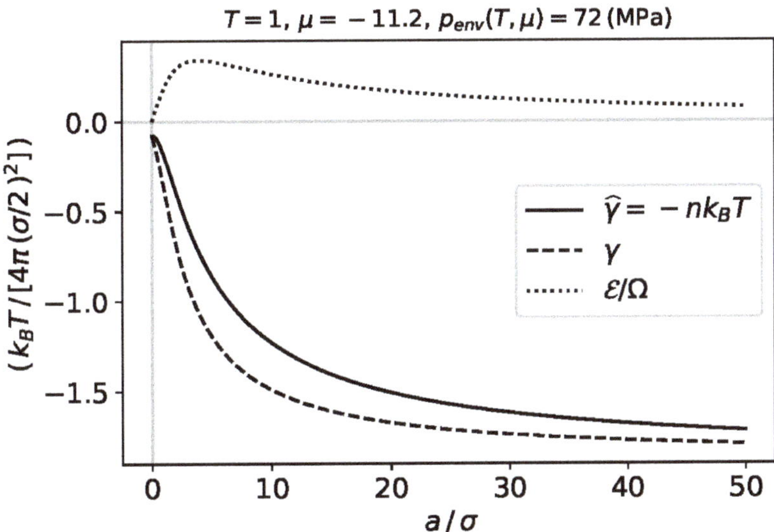

Figure 2. Adsorbed phase thermodynamic quantities per unit area at constant temperature and chemical potential. The figure shows that when the system is small the characteristic energy of the adsorbed phase per unit area depends on the size of the adsorbent. The transition from small to macroscopic is continuous and may reasonably be considered to be beyond 50 σ for this system.

Figure 3 shows the characteristic energy per molecule, the energy $\gamma\Omega$ per molecule, and the subdivision potential per molecule $\mathcal{E}/N$, as functions of the adsorbent radius a at constant temperature and chemical potential. We observe that when the system becomes larger, $\mathcal{E}$ approaches zero, and γ approaches the limit value $-k_B T$. Here, the characteristic energy per molecule follows the 2-dimensional analogue of the ideal gas law $\widehat{\gamma}\Omega = -nk_B T$.

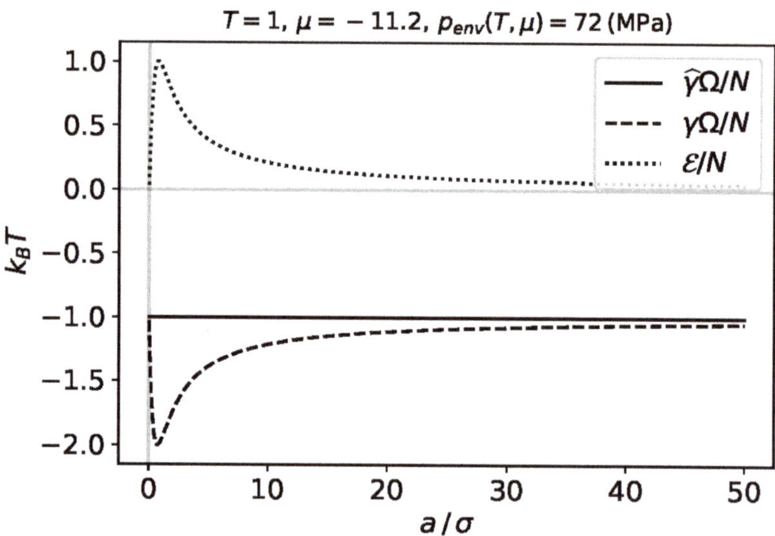

Figure 3. Adsorbed phase thermodynamic quantities per molecule at constant temperature and chemical potential.

7. Discussion

As we are considering a two-dimensional ideal gas, or a dilute adsorbed phase with free mobility, we expect the phase to follow the two-dimensional analogue of the ideal gas law $\widehat{\gamma} = -nk_BT$. This is consistent with Equation (23) and Figure 3. When the adsorbent size a approaches zero, the dividing surface radius R should approach the potential minimum distance of the interaction between a single adsorbent atom and an adsorbed molecule. This distance is $2^{1/6}\sigma$ for the Lennard-Jones potential. This is because of the way we have defined R.

In the macroscopic limit the energy $\gamma\Omega$ is a linear function of Ω. The tension γ then becomes equal to the characteristic energy per unit area of the adsorbed phase, i.e., $\gamma = (U - TS - \mu N)/\Omega$, for the environment T, Ω, μ. When the system is small $\gamma\Omega$ deviates from the characteristic energy by $\mathcal{E}$, as shown by Equation (12). The integral surface tension was therefore defined to represent this important quantity, i.e., $\widehat{\gamma} = (U - TS - \mu N)/\Omega = X/\Omega$. The tension γ is now given by the relation $\gamma = \partial X/\partial\Omega$. This relation is always valid whether the system is small or large. It is only in the special case of the macroscopic limit that the relation $\partial X/\partial\Omega = X/\Omega$ is true. The integral and differential surface tensions are then equal. This way of stating the smallness is expressed by Equations (9) and (10), which may be rewritten as

$$\frac{\widehat{\gamma}\Omega}{\Omega} - \left(\frac{\partial\widehat{\gamma}\Omega}{\partial\Omega}\right)_{T,\mu} = \frac{X}{\Omega} - \left(\frac{\partial X}{\partial\Omega}\right)_{T,\mu} = \frac{\mathcal{E}}{\Omega} \tag{36}$$

where $\mathcal{E}$ goes to zero in the macroscopic limit. Both $\widehat{\gamma}$ and γ are functions of the system size, thus $\mathcal{E}$ measures the difference expressed by Equation (36), and not simply the difference between the characteristic energy and the limit of $\gamma\Omega$.

The effects of size on intensive variables, characteristic of small systems, may be expressed by the subdivision potential $\mathcal{E}$ and its derivatives, according to Equation (14). The physical significance of $\mathcal{E}$ is more clear if we use the definitions $\mathcal{E} \equiv (\widehat{\gamma} - \gamma)\Omega = X - \gamma\Omega$ and $\Omega_t \equiv \mathcal{N}\Omega$ to rewrite Equation (1) as

$$dU_t = T\,dS_t + \gamma\,d\Omega_t + \mu\,dN_t + \mathcal{E}\,d\mathcal{N} \tag{37}$$

An alternative definition of $\mathcal{E}$ is then

$$\mathcal{E} \equiv \left(\frac{\partial U_t}{\partial \mathcal{N}}\right)_{S_t,\Omega_t,N_t} = -\frac{\Omega}{\mathcal{N}}\left(\frac{\partial U_t}{\partial \Omega}\right)_{S_t,\Omega_t,N_t} \tag{38}$$

By this definition, we see that $\mathcal{E}$ is the work required to increase the number of replicas while keeping S_t, Ω_t, and N_t constant. Since the total surface area is constant, it must be redistributed across the new number of replicas. The area of each replica therefore becomes smaller, which for a fixed shape means larger curvature. Thus, the subdivision potential for the given system, with fixed shape, is also the work required to change the adsorbed phase curvature while keeping S_t, Ω_t, and N_t constant. If the process of adding a system to the ensemble is at constant Ω instead of Ω_t the work is given by $X = \hat{\gamma}\Omega = \gamma\Omega + \mathcal{E}$.

When the intensive properties become independent of the curvature $\mathcal{E} = 0$, which is consistent with Equations (9) and (10). This occurs in the macroscopic limit, when the adsorbent becomes large, which is consistent with Figures 2 and 3. All the differential coefficients, expressing dependence of intensive properties on curvature, are then zero, and $d\mathcal{E} = 0$ by Equation (14). It also follows from these figures that a must be larger than 50σ for $\mathcal{E}$ to become small.

8. Concluding Remarks

The above analysis shows that we can use the adsorbed phase as a small thermodynamic system in the sense of Hill. The analysis for our ideal adsorbed gas model becomes very simple. This allows the close relationship between the subdivision potential and the dependence of intensive properties on size, and the internal structure of nanothermodynamics to be seen more clearly.

Author Contributions: All authors have contributed equally to the work. Conceptualization, B.A.S., D.B. and S.K.S.; Formal analysis, B.A.S., D.B. and S.K.S.; Investigation, B.A.S., D.B. and S.K.S.; Methodology, B.A.S., D.B. and S.K.S.; Writing—original draft, B.A.S., D.B. and S.K.S.; Writing—review & editing, B.A.S., D.B. and S.K.S. All authors have read and agreed to the published version of the manuscript.

Funding: B.A.S. and S.K.S. acknowledge funding by the Research Council of Norway via project number 275754, and NTNU through the Outstanding Academic Fellows program. D.B. acknowledges funding by the Research Council of Norway through its Center of Excellence Funding Scheme, project number 262644, PoreLab.

Conflicts of Interest: The authors declare no conflict of interest.

Appendix A

The Gibbs-Duhem relation for a single component homogeneous phase α is given by

$$d\mu^\alpha = -s^\alpha\, dT^\alpha + v^\alpha\, dp^\alpha \tag{A1}$$

where $s^\alpha \equiv S^\alpha/N^\alpha$ is the entropy per molecule, and $v^\alpha \equiv V^\alpha/N^\alpha$ is the volume per molecule. At chemical equilibrium the chemical potential μ of the adsorbed phase, and the chemical potential μ^G of the gas phase are subject to the relations $\mu = \mu^G$ and $d\mu = d\mu^G$. The Gibbs-Duhem relation for the ideal gas may then be written as

$$d\mu = -s^G\, dT + kT\, d\log p \tag{A2}$$

where p is the equilibrium gas pressure, T is the equilibrium temperature, and superscript G denotes the gas phase.

Using Equations (6), (8) and (A2) we have

$$-(dX)/N - (S/N)\,dT + (\gamma/N)\,d\Omega = -s^G\,dT + kT\,d\log p$$

$$\frac{S}{N} - s^G = -kT\left(\frac{\partial \log p}{\partial T}\right)_{X,\Omega} \tag{A3}$$

$$= -kT\left(\frac{\partial \log p}{\partial T}\right)_{\hat{\gamma},\Omega}$$

From Equation (8) and the definition $F \equiv \hat{\gamma}\Omega + \mu N$ we have

$$dF = -S\,dT + \gamma\,d\Omega + \mu\,dN \tag{A4}$$

From this equation we obtain the Maxwell relation

$$\left(\frac{\partial \mu}{\partial T}\right)_{\Omega,N} = -\left(\frac{\partial S}{\partial N}\right)_{T,\Omega} \tag{A5}$$

For variations in the adsorbed phase chemical potential at constant Ω and N, at equilibrium with the gas, we then have

$$d\mu = \left(\frac{\partial \mu}{\partial T}\right)_{\Omega,N}dT = -\left(\frac{\partial S}{\partial N}\right)_{T,\Omega}dT = -s^G\,dT + kT\,d\log p$$

$$\left(\frac{\partial S}{\partial N}\right)_{T,\Omega} - s^G = -kT\left(\frac{\partial \log p}{\partial T}\right)_{\Omega,N} \tag{A6}$$

From Equation (6) and the equilibrium condition $\mu = \mu^G$ we have

$$\frac{U - TS - X}{N} = \frac{U^G - TS^G + pV^G}{N^G}$$

$$= u^G - Ts^G + kT \tag{A7}$$

where $u^G \equiv U^G/N^G$ is the gas internal energy per gas molecule. We also have from Equation (8), using Equation (A7) to eliminate $X/(TN)$:

$$d\mu = -\frac{1}{N}\,dX - \frac{S}{N}\,dT + \frac{\gamma}{N}\,d\Omega$$

$$= -\frac{1}{N}\,d(TX/T) - \frac{S}{N}\,dT + \frac{\gamma}{N}\,d\Omega$$

$$= -\frac{T}{N}\,d(X/T) - \left(\frac{X}{TN} + \frac{S}{N}\right)dT + \frac{\gamma}{N}\,d\Omega \tag{A8}$$

$$= -\left(\frac{X}{TN} + \frac{S}{N}\right)dT, \quad \left(\frac{X}{T},\Omega\,\text{const.}\right)$$

$$= \frac{u^G - Ts^G + kT - U/N}{T}\,dT, \quad \left(\frac{X}{T},\Omega\,\text{const.}\right)$$

Using $d\mu = d\mu^G$ and Equation (A8) we obtain

$$\frac{u^G - Ts^G + kT - U/N}{T}\,dT = -s^G\,dT + kT\,d\log p$$

$$\frac{U}{N} - u^G = kT\left[1 - T\left(\frac{\partial \log p}{\partial T}\right)_{\frac{X}{T},\Omega}\right] \tag{A9}$$

Using Equations (7), (A2), (A4) and (A7) we have

$$-\left(\frac{\partial S}{\partial N}\right)_{T,\Omega} = -\frac{1}{T}\left(\frac{\partial U}{\partial N}\right)_{T,\Omega} + \frac{\mu}{T}$$

$$= -\frac{1}{T}\left(\frac{\partial U}{\partial N}\right)_{T,\Omega} + \frac{u^G - Ts^G + kT}{T}$$

$$-s^G\,dT + kT\,d\log p = \left[-\frac{1}{T}\left(\frac{\partial U}{\partial N}\right)_{T,\Omega} + \frac{u^G - Ts^G + kT}{T}\right]dT \tag{A10}$$

$$kT\left(\frac{\partial \log p}{\partial T}\right)_{\Omega,N} = -\frac{1}{T}\left[\left(\frac{\partial U}{\partial N}\right)_{T,\Omega} - u^G - kT\right]$$

$$\left(\frac{\partial U}{\partial N}\right)_{T,\Omega} - u^G = kT\left[1 - T\left(\frac{\partial \log p}{\partial T}\right)_{\Omega,N}\right]$$

References

1. Hill, T.L. Thermodynamics of Small Systems. *J. Chem. Phys.* **1962**, *36*, 3182. [CrossRef]
2. Hill, T.L. *Thermodynamics of Small Systems*; Dover Books on Chemistry; Dover Publications: Mineola, NY, USA, 2013.
3. Bedeaux, D.; Kjelstrup, S.; Schnell, S.K. *Nanothermodynamics—General Theory*; PoreLab: Trondheim, Norway, 2020.
4. Bedeaux, D.; Kjelstrup, S. Hill's nano-thermodynamics is equivalent with Gibbs' thermodynamics for surfaces of constant curvatures. *Chem. Phys. Lett.* **2018**, *707*, 40–43. [CrossRef]
5. Strøm, B.A.; Simon, J.M.; Schnell, S.K.; Kjelstrup, S.; He, J.; Bedeaux, D. Size and shape effects on the thermodynamic properties of nanoscale volumes of water. *Phys. Chem. Chem. Phys.* **2017**, *19*, 9016–9027. [CrossRef] [PubMed]
6. Hill, T. *Statistical Mechanics: Principle and Selected Applications*; McGraw-Hill Series in Advanced Chemistry; McGraw-Hilll: New York, NY, USA, 1956.
7. McQuarrie, D.; A, M. *Statistical Mechanics*; Chemistry Series; Harper & Row: New York, NY, USA, 1975.
8. Hill, T.L. Statistical Mechanics of Adsorption. V. Thermodynamics and Heat of Adsorption. *J. Chem. Phys.* **1949**, *17*, 520–535. [CrossRef]
9. Hill, T.L. Statistical Mechanics of Adsorption. IX. Adsorption Thermodynamics and Solution Thermodynamics. *J. Chem. Phys.* **1950**, *18*, 246–256. [CrossRef]

 *nanomaterials*

Article

Thermodynamics of Adsorbed Methane Storage Systems Based on Peat-Derived Activated Carbons

Ilya Men'shchikov *, Andrey Shkolin, Elena Khozina and Anatoly Fomkin

Dubinin Laboratory of Sorption Processes, Frumkin Institute of Physical Chemistry and Electrochemistry, Russian Academy of Sciences, Leninskii Prospect 31, Building 4, 119071 Moscow, Russia; shkolin@bk.ru (A.S.); khozinaelena@gmail.com (E.K.); fomkinaa@mail.ru (A.F.)

* Correspondence: i.menshchikov@gmail.com; Tel.: +7-(495)-952-85-51

Received: 22 June 2020; Accepted: 13 July 2020; Published: 15 July 2020

Abstract: Two activated carbons (ACs) were prepared from peat using thermochemical K_2SO_4 activation at 1053–1133 K for 1 h, and steam activation at 1173 K for 30 (AC-4) and 45 (AC-6) min. The steam activation duration affected the microporous structure and chemical composition of ACs, which are crucial for their adsorption performance in the methane storage technique. AC-6 displays a higher micropore volume (0.60 cm^3/g), specific BET surface (1334 m^2/g), and a lower fraction of mesopores calculated from the benzene vapor adsorption/desorption isotherms at 293 K. Scanning electron microscopy (SEM), X-ray diffraction (XRD), and small-angle X-ray scattering (SAXS) investigations of ACs revealed their heterogeneous morphology and chemical composition determined by the precursor and activation conditions. A thermodynamic analysis of methane adsorption at pressures up to 25 MPa and temperatures from 178 to 360 K extended to impacts of the nonideality of a gaseous phase and non-inertness of an adsorbent made it possible to evaluate the heat effects and thermodynamic state functions in the methane-AC adsorption systems. At 270 K and methane adsorption value of ~8 mmol/g, the isosteric heat capacity of the methane-AC-4 system exceeded by ~45% that evaluated for the methane-AC-6 system. The higher micropore volume and structural heterogeneity of the more activated AC-6 compared to AC-4 determine its superior methane adsorption performance.

Keywords: activated carbon; high-pressure methane adsorption; thermodynamics of adsorption systems

1. Introduction

Storage of adsorbed natural gas (ANG) represents an effective alternative to the known techniques of liquefied and compressed gas due to several reasons, such as high specific capacity, safety, and energy efficiency [1–3]. In most studies, when choosing an adsorbent for an ANG system, the most crucial criterion is the methane adsorption capacity [2–4]. Among adsorbents, activated carbons (AC) [2,4–8], metal-organic frameworks (MOF) [5,9,10], and porous organic polymers (POP) [10,11] are the most promising for use in ANG technology because of their high adsorption capacity of methane, which is the major component of natural gas. It should be noted that the utilization of these materials for the ANG storage systems implies that they meet the requirements imposed on these systems. Among these requirements are chemical stability, mechanical strength, thermal stability, conductivity, minimized affinity to water, and heavier natural gas components (ethane, propane, and butane), accessibility and simplicity of the synthesis, and low costs of reactants and final product [2,9]. Although MOFs show high promise in this area, their use in the loose powder form with low packing density produced by most conventional synthetic techniques is not practical in the ANG systems, and, therefore, they must be shaped [12,13]. Shaping or densification of MOFs may cause degradation of their porous structure due to their low resistance against mechanical loads [14], which leads to a deterioration in the methane adsorption capacity [15]. Moreover, Zhang et al. [16] found that the most promising

HKUST-1 adsorbent lost 32% of its adsorption capacity over 200 sorption/desorption cycles due to the irreversible blocking of almost half of the pore volume. However, despite the advances in developing new synthetic technologies, the large-scale production of MOF- and POP-based adsorbents is limited by a low level of reproducibility between different batches of final product and high costs of the production process [10,12,17].

Therefore, for the foreseeable future, microporous ACs with their unique properties, such as high thermal stability and mechanical strength [2,5,8], remain most suitable for the ANG storage technique. Additional properties of ACs, which are essential for the ANG systems, are the high adsorption capacity per unit volume, preferential adsorption of gas in the presence of moisture (hydrophobicity), low resistance to gas flow, and complete release of adsorbates at increasing temperatures and decreasing pressures [18]. As shown in the experimental study of methane adsorption in a series of commercial (Maxsorb, Filtrasorb 400, NuCharRGC300) and laboratory synthesized ACs with various pore size distribution and MOF (HKUST-1), a properly designed carbon adsorbent with an optimal micropore/mesopore ratio demonstrates an excellent adsorption performance at high pressures without any loss in the porosity [5].

When searching for an efficient carbon adsorbent for methane storage, an important criterion is the availability of raw materials. Peat, matching this criterion, is a partially carbonized phytomass (up to 61%), and it is considered a suitable raw material for the production of carbonaceous adsorbents in terms of energy consumption and lows cost [19,20]. As a fossil raw material with an extensive capacity, peat finds use as a precursor for the synthesis of promising adsorbents for many applications, including ANG technology. From this standpoint, it is crucial to determine the influence of both raw material and activation conditions on the porous structure, chemical composition, and morphology of the final product—activated carbon—and its adsorption properties relative to methane. The adsorption performance of a porous material with respect to a gas can be predicted based on the structural and energy characteristics (SEC) calculated from data on standard vapor adsorption according to the Dubinin theory of volume filling of micropores (TVFM) [21]. Therefore, it is essential to establish a relationship between the methane adsorption capacity of peat-derived adsorbents and their SEC values, which, in turn, are affected by chemical composition and morphology inherited from peat and methods of their production [22].

It should also be noted that a proper design of ANG storage facilities operating within a wide range of temperatures and at high pressures requires accounting for both the heat release/absorption and deformation of an adsorbent upon the adsorption/desorption processes. These phenomena have a significant impact on the thermodynamic characteristics of gas adsorption systems at high pressures [23]. Many authors calculated the heat of adsorption by substituting various models of adsorption [24,25] into the well-known Clausius–Clapeyron equation, in which an adsorbate in its gaseous phase is endowed with ideal gas-phase behaviors.

An approach for calculating the thermodynamic functions of adsorption systems, which takes into account the effects associated with the nonideality of an equilibrium gas phase and non-inertness of an adsorbent at high pressures, has been developed by Bakaev [26,27]. Behind this approach is the concept by Guggenheim that the thermodynamics of the adsorption equilibrium is rigorous if the thermodynamic functions of an adsorption system are evaluated using only the quantities, which are measured experimentally [28–30]. Using the methods of replacement of variables and Jacobian-determinant in thermodynamics of adsorption equilibrium, Bakaev derived an equation, which is recognized as a complete formula for calculating the differential molar isosteric heat of adsorption [26]. This approach makes it possible to neglect the surface effects and avoid the difficulty in interpreting adsorption data associated with the Gibbsian treatment of adsorption phenomena in terms of interfacial excess amounts [31].

The applicability of the Bakaev equation was demonstrated by calculating the thermodynamic functions characterizing the adsorption of gases (Xe, Kr, Ar, N_2, O_2, H_2, CH_4, CO_2, and He) in microporous adsorbents (zeolites and ACs) in a wide range of temperature and pressure [32–34]. It should also be noted that an analysis of thermodynamic parameters of an adsorption system, as a function of pressure, temperature, and amounts of adsorbed methane, also provides information on changes in the state of methane molecules in micropores [32]. Indeed, in [5], an advantage in the performance of ACs in terms of the working capacity compared to HKUST-1 was attributed to a lower isosteric heat of adsorption at the early adsorption stages.

The objective of the present work is to evaluate and compare the SEC values of two ACs, which were produced from the same raw material as peat, but at different activation durations, from the standard adsorption measurements by TVFM equations. The second task is to find a correlation between the SEC values of the AC samples and data on their chemical composition and morphology obtained by independent structural methods as X-ray diffraction (XRD), small-angle X-ray scattering (SAXS), and scanning electron microscopy. Our study also focuses on the thermodynamic functions derived qualitatively for methane adsorption in two peat-derived ACs measured at pressures up to 25 MPa and over the temperatures from 178 to 360 K, i.e., both in sub- and supercritical states. The data are essential for identifying the optimal thermal conditions of such a system, including the charge/delivery processes.

2. Materials and Methods

2.1. Adsorbent

Shredded high-grade metamorphic peat from Central Russia (supplied by JSC Electrostal Research and Production Association "Neorganika", Electrostal, Russia) was used as a precursor for the production of the activated carbons in three stages, including carbonization, thermochemical activation, and steam activation.

After passing the preliminary treatment of crushing and separation, peat was impregnated with potassium sulfide (K_2SO_4, > 99.0%, Sigma-Aldrich, St. Louis, MO, USA) and mixed to form a homogeneous elastic paste. Afterward, the paste preform was pressed and granulated with the use of extruding jets. The cylindrical granules thus obtained were dried and carbonized in a drum furnace at the temperatures of 873–973 K in order to remove both the residual water and volatile compounds, and decompose the peat organic matter.

Then the granulated material was activated in a rotary furnace at the temperature within a range of 1053–1133 K for 1 h. At this stage, the volatile compounds were completely removed. As a result, the material is enriched with carbon, which provides its mechanical strength and density. The activation procedure gives rise to a porous structure in the carbonizate [20]. Due to the peculiarities of the production process of carbon adsorbents from peat with the complex chemical composition, the activated granules can contain up to 30 wt.% potassium, 10% sulfur, and up to 10 wt.% ash components [35].

For this reason, the activated granules were cooled, and then they passed through a special washing series including leaching with returned alkali solutions, washing in water and hot hydrochloric acid (EKOS 1, Staraya Kupavna, Moscow Region, Russia), and a final rinsing with water. In the end, the wet granules were placed into a tumble drier to reduce moisture content up to 3–5 wt.%.

As follows from [19,20,36], the optimal temperature of heat treatment of carbon-containing biomass ranges from 1073 to 1173 K. In order to develop microporosity, the granules were subjected to additional steam activation carried out in a reactor at 1173 K. It is known that at a particular temperature of activation, weight loss due to burn-off (or degree of activation) increases with activation duration producing micropores up to excessive activation, resulting in an increase in transport pores and macropores [36,37] and a decrease in the characteristic energy of adsorption [21]. According to [38,39], the burn-off of 30–40% matches the development of a maximum microporosity in activated carbons. We attained these values at the activation times varied from 30 to 45 min. By using two different activation duration, 30 and 45 min, we obtained two peat-based AC samples labeled as AC-4 and AC-6, respectively. Due to the shorter duration of this stage, AC-4 had a lower degree of activation compared to AC-6.

2.2. Adsorptive

The adsorptive gas used in the experiments was high purity (99.999%) methane. Methane has the following physicochemical properties: molecular mass M = 16.0426 g/mol; boiling temperature T_0 = 111.66 K; critical temperature T_{cr} = 190.77 K; critical pressure P_{cr} = 4.641 MPa [40].

2.3. Methods

Porous characteristics of the activated carbons were estimated from the isotherms of benzene standard vapor adsorption at 293 K measured by employing an original gravimetric vacuum adsorption setup designed in IPCE RAS [41,42]. A maximal measurement error was ± 0.19 mmol/g with a confidence level of 99% determined according to [43]. A LOIP LT-411 thermostatic bath circulator and a LOIP FT-311-80 cryostat (ultralow temperatures) were used for temperature control with an error of ± 0.01 K. The SEC values of the carbons, such as a specific volume of micropores W_0, standard characteristic energy of adsorption E_0, and effective half-width of micropores x_0, were evaluated by applying the Dubinin–Radushkevich (D-R) equation [21]. The Brunauer–Emmett–Teller (BET) method [44] was used to calculate specific BET surface S_{BET} from the data on benzene adsorption at 293 K with a benzene molecular area of 0.40 nm^2 for a flat molecular orientation in a monolayer [45]. The specific surface area of mesopores (S_{meso}) was calculated from the benzene adsorption/desorption data using the well-known Kiselev equation [46]. The specific mesopore volume was calculated as $W_{meso} = W_S - W_0$, where W_S is the total pore volume calculated from the amount of benzene adsorption at the relative pressure P/P_s = 0.99.

The surface morphology and elemental composition of mechanically ground AC-4 and AC-6 were examined by scanning electron microscopy (SEM) using a Quanta 650 FEG microscope (FEI Company, Hillsboro, OR, USA) equipped with an Oxford Energy Dispersive X-ray (EDX) detector operating at 30 kV accelerating voltage. The data on the elemental composition of both samples were obtained by averaging ten measurements.

The phase composition of AC-4 and AC-6 was examined by powder X-ray diffraction (XRD) with an Empyrean Panalytical (Panalytical BV) diffractometer in Bragg–Brentano geometry using Ni-filtered CuKα-radiation (λ_{Cu} = 0.1542 nm) in the 2θ angular range of 10–120 degrees. The samples were ground to powder, and no binder was employed. The ICDD PDF2 database was used for phase identification. The detailed qualitative analysis was carried out through the graphite characteristic reflections—(002), (10), (100), (101), and (11).

The results from benzene adsorption analysis based on TVFM were compared with SAXS data on the porous structure of ACs obtained at a SAXSess diffractometer (Anton Paar GmbH, Graz, Austria). Powdered samples were measured in transmission geometry in a vacuum chamber. Monochromatic Cu-Kα radiation and a 2D Imaging Plate detector were employed, and the range of scattering q vectors was from 0.1 to 27 nm^{-1}.

Methane adsorption equilibria on the AC-4 and AC-6 adsorbents were studied within a range of pressures from 5 Pa to 25 MPa and at the temperatures varied from 178 to 360 K by the volumetric-gravimetric method using three original adsorption devices developed in IPCE RAS and described in detail in our previous works [47–49]:

- a semi-automatic adsorption weight vacuum unit (from 5 Pa to 0.1 MPa, gravimetric method; accuracy of ± 1.5%) [47];
- a universal adsorption-dilatometer setup (0.1–6 MPa, volumetric method, accuracy of ± 3%) [48];
- an original volumetric-gravimetric high-pressure set-up (0.2–25 MPa, accuracy ± 5%) [49].

The values of methane adsorption a were determined as the total content of adsorbed methane in micropores of an adsorbent:

$$a = (V - V_a)\rho_g / m_0 \tag{1}$$

where V is the total geometric volume of the system, V_a is the volume of an adsorbent with micropores, ρ_g is the density of gaseous phase at given pressure P and temperature T, m_0 is the mass of a regenerated adsorbent. As follows from the data below, the specific surface of mesopores in both adsorbents constituted no more than 10% of the total specific surface, and therefore, one can neglect the adsorption processes in mesopores. When calculating the value of methane adsorption a, the volume of adsorbent with micropores was determined as a sum of the volume of this adsorbent determined via helium pycnometry, V_{He}, and product $m_0 \cdot W_0$, where micropore volume W_0 is evaluated from the data on benzene adsorption at 293 K using the D-R equation [21].

3. Results and Discussion

3.1. Structure and Morphology Characterization

In order to reveal the influence of the activation time on the porous structure of the peat-derived carbon adsorbents, the SEC values and the other parameters characterizing mesopores for AC-4 and AC-6 summarized in Table 1 should be compared.

Table 1. The parameters of the porous structure of the peat-derived activated carbons.

Sample	E_0, kJ/mol	W_0, cm^3/g	x_0, nm	S_{BET}, m^2/g	W_t, cm^3/g	S_{meso}, m^2/g	W_{meso}, cm^3/g
AC-4	20.6	0.48	0.58	957	0.72	105	0.24
AC-6	19.1	0.60	0.63	1334	0.70	96	0.10

As follows from Table 1, the steam activation duration had a significant impact for developing the microporosity. It should be noted that ceteris paribus, the additional time of activation of AC-6 was responsible for the development of a specific BET surface through the enlargement of micropores without the formation of a noticeable amount of mesopores. The AC-4 sample is characterized both by the higher energy of adsorption and the proportion of mesopores in the total pore volume compared to AC-6. One can expect that the differences in the porous structure of these two ACs manifest in variations in the methane adsorption capacities.

The SEM images (Figure 1a,b) allowed for a qualitative assessment of the surface morphology of AC-4 and AC-6. The images indicate that the carbons are not homogeneous in density. As can be seen from the comparison of Figure 1a,b, the heterogeneous surfaces of the AC-4 and AC-6 powders are characterized by different proportions of relatively dark (dense carbon phase) and light (relatively low-dense carbon phase) areas. The steam activation stage implies a partial burning-off of a non-graphite amorphous phase associated with the light areas in the SEM images. Therefore, the SEM image of more activated AC-6 shows fewer light inclusions compared to AC-4.

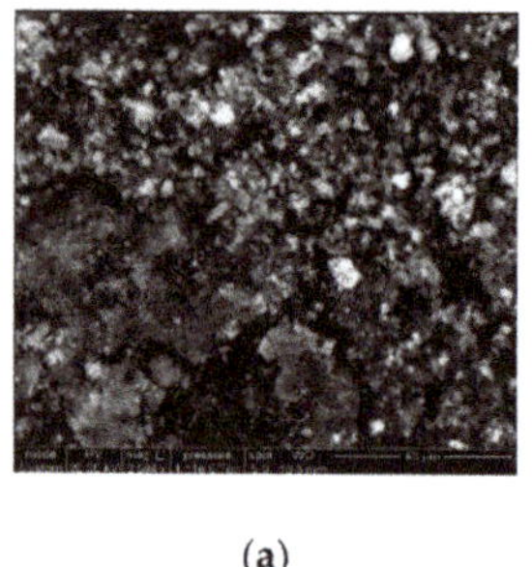

(a)   (b)

Figure 1. Scanning electron microscopy (SEM) images of the peat-derived AC-4 (**a**) and AC-6 (**b**) adsorbents. The scale bar is 40 μm.

The heterogeneous surface morphology of the peat-based adsorbents correlates with the SEM-EDX data on their elemental chemical composition given in Table 2. The diverse chemical composition of the adsorbents is reliant on the precursor (peat), activation process, and activating agent (K_2SO_4). Therefore, besides the relatively high percentage of oxygen, especially in AC-6 (see Table 2), the chemical composition of both carbon adsorbents also includes such heteroatoms as Si, Al, Ca, Fe, the presence of which can be attributed to ash and plant residues decomposed under peat-forming geochemical processes [35]. Potassium and sulfur are residuals of the activating agent.

Table 2. Elemental chemical composition of the peat-derived activated carbons.

Sample	Elements (wt.%)										
	C	O	K	S	Si	Al	Cl	Ca	Fe	Mg	B
AC-4	74.9	10.7	0.8	3.0	2.4	0.6	0.6	1.4	0.3	0.18	4.9
AC-6	60.8	23.7	0.5	1.7	10.2	1.2	0.6	1.3	-	-	-

As seen from Table 2, the percentages of carbon in both peat-based adsorbents are less than that found for commercial peat-based Sorbonorit 4 (Norit Ltd., Netherland): 95 wt.% [50], and close to the data for the peat-based ACs prepared via a one-step steam activation process (64–91 wt.%) reported in [51].

When comparing the data on the elemental chemical composition of AC-4 and AC-6 in Table 2, one can conclude that the increase in the steam activation time reduced the relative amounts of carbon, potassium, and sulfur. In AC-6, metal impurities such as Fe and Mg were not found. At the same time, the percentage of unremovable Si and O increased. Thus, the higher activation time of AC-6 caused the higher oxygen/carbon (O/C) ratio: ~0.39 compared to AC-4 (O/C = 0.14).

The presence of heteroatoms (O, K, S, etc.), which form surface functional groups, largely determines the surface chemical characteristics of the carbon adsorbents, thereby affecting their interactions with adsorptive, and as a consequence, the adsorption process, especially in the early stages of adsorption.

The analysis of powder XRD patterns for both samples (see Figure 2) revealed the pronounced reflections related to carbon species in hexagonal graphite lattice and quartz. The graphite crystallites are likely inherited from peat with a variable degree of thermal metamorphism.

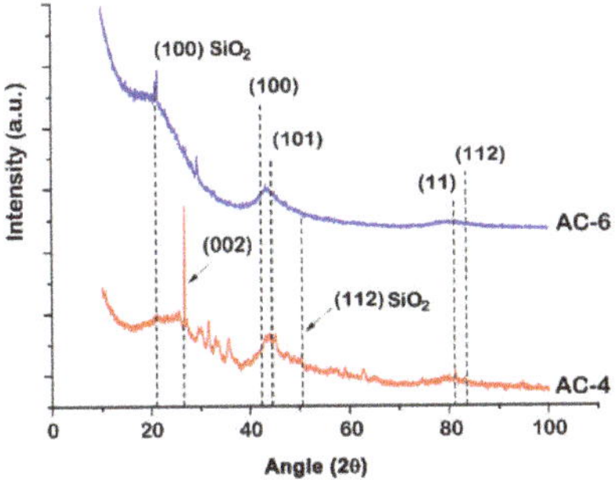

Figure 2. X-ray diffraction (XRD) patterns for the peat-derived activated carbons: AC-4 and AC-6. Dashed lines show the primary reflections of graphite and quartz.

The activation stages promoted the growth and subsequent ordering of the crystalline phases to a certain limit. But over this limit of activation, the peak shapes are somewhat broadened, and their intensity is lowered (compare AC-4 and AC-6), which is indicative of a more amorphous and non-graphitized structure in AC-6. Moreover, the strength of the diffraction intensity of AC-6 at low angles is more significant compared to that of AC-4, which is consistent with a higher density of pores in AC-6 (see Table 1) [52].

As shown in [53,54], SAXS is able to provide topological information from the molecular to mesopore scale, regardless of the sample crystallinity. Characterization of the carbons by SAXS is shown in Figure 3.

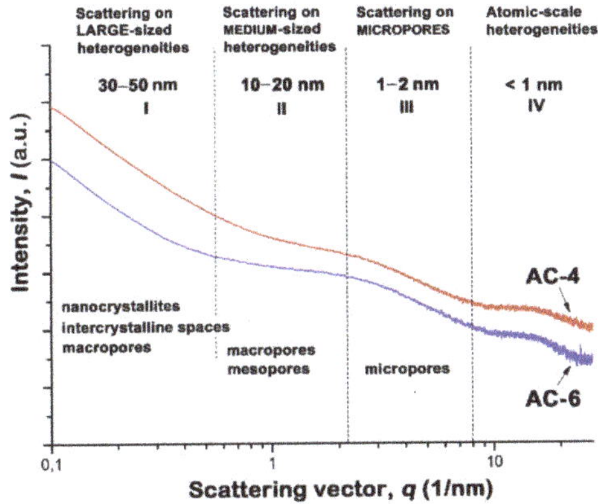

Figure 3. The log-log plot of scattering intensities versus scattering vector q for the peat-derived activated carbons obtained at different steam activation times.

The linear log-log Guinier plots of the scattering intensity I on the scattering vector q in the region between 2 and 8 nm^{-1} (marked as III in Figure 3) indicate scattering by a monodisperse system [53,54]. Therefore, the values of radius of gyration used for estimations of pore sizes, R_G, can be calculated with the use of the tangent method and Guinier formula for a slit-like model of pores [55]: $q^2 I(q) = I_0 \cdot \exp\left[-R_G^2 q^2\right]$. We found that the values of R_G for AC-4 and AC-6 are close enough to each other: 0.48 and 0.50 nm, respectively.

When analyzing the experimental SAXS data for the carbons under study (see Figure 3), we need to compare them with an empirical relationship proposed by Dubinin with coworkers for carbon adsorbents with relatively homogeneous chemical composition and narrow slit-like pore size distribution [56–58]:

$$E_0 R_G = 14.8 \pm 0.6 \ \text{kJ·nm/mol} \tag{2}$$

where E_0 is the characteristic energy of adsorption determined from the adsorption of standard benzene vapors (see Table 1) and R_G is the radius of gyration of micropores evaluated from the SAXS data.

We found that the values of $E_0 \cdot R_G$ determined for AC-4 and AC-6 (10.3 and 9.1 kJ·nm/mol, respectively) differ markedly from that in Equation (2). Since the magnitudes of E_0 for AC-4 and AC-6 from the adsorption data are typical for microporous adsorbents [see, for example, Reference [32]], we attribute this deviation to an effect of the multi-component chemical composition of peat on the formation of turbostratic carbon nanocrystallites in the final ACs. As a consequence, the variation in the sizes and shapes of carbon nanocrystallites over the sample plus the amorphous phase detected by XRD made it impossible to describe the porous system using a particular model of pore shapes for correct evaluation of gyration radius R_G from the SAXS data. Thus, the surface chemistry of ACs can be a crucial feature of novel peat-based adsorbents and should be taken into account when developing new adsorption systems. Undoubtedly, the structural data only for two samples are insufficient to establish an unambiguous correlation between the SEC quantities and real structural properties of ACs. Nevertheless, these new findings contribute to establishing a general pattern needed for the synthesis of adsorbents with tailored structure.

3.2. Methane Adsorption on the Peat-Derived Carbon Adsorbents

Figure 4 presents experimental isotherms of methane adsorption in AC-4 and AC-6 at the temperatures varied from 178 to 360 K and pressures up to 25 MPa.

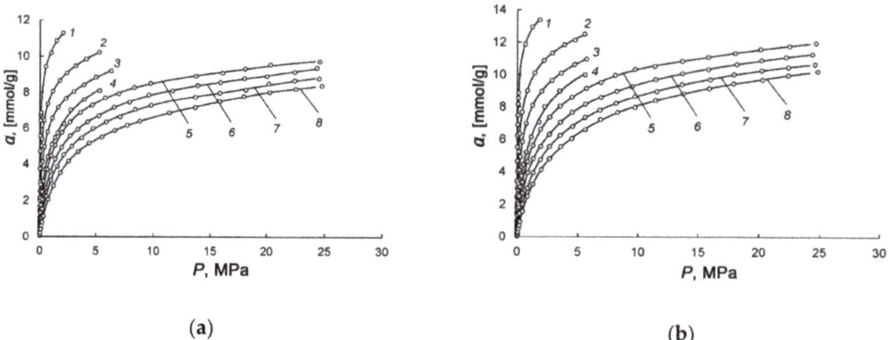

(a)   (b)

Figure 4. Dependences of methane adsorption in AC-4 (**a**) and AC-6 (**b**) on pressure at temperatures, K: 178 (*1*), 216 (*2*), 243 (*3*), 273.15 (*4*), 300 (*5*), 320 (*6*), 340 (*7*); 360 (*8*). Experimental data are shown by symbols, and lines are the approximation of experimental data by Equation (3).

We used a formula derived by Bakaev for the total adsorption isotherm [59] to fit the experimental data shown in Figure 4a,b:

$$a(P) = \frac{k_0 \left(k_1 P + 2k_2 P^2 + 3k_3 P^3\right)}{1 + k_1 P + k_2 P^2 + k_3 P^3} \tag{3}$$

where k_0 characterizes an adsorption system, k_1, k_2, k_3 are the temperature-dependent and numerically adjusted coefficients, and P is the equilibrium pressure expressed in Pa.

Equation (3) was successfully used for describing the adsorption process in various adsorbents, including zeolites [59], polymer sorbents [60], and ACs [61]. In our case, the maximum regression error did not exceed 3%, which made it possible to calculate a set of adsorption and thermodynamic parameters of the systems with high accuracy.

It can be seen from Figure 4a,b that over the entire range of temperatures and pressures, the more activated AC-6 with a large volume fraction of micropores shows a higher adsorption capacity for methane compared to AC-4.

Figure 5a,b demonstrate the isosteres of methane adsorption in AC-4 and AC-6, which were calculated from the isotherm of methane adsorption shown in Figure 4a,b.

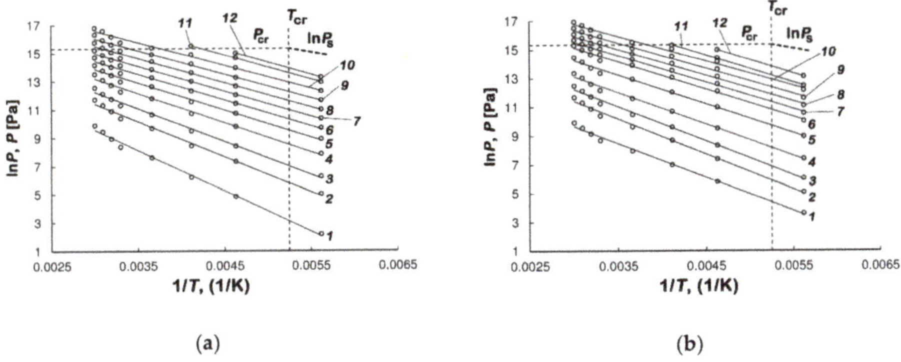

(a)　　　　　　　　　　　　　　　　**(b)**

Figure 5. The isosters of methane adsorption in AC-4 (**a**) and AC-6 (**b**) at the values of methane adsorption, mmol/g: 0.1 (*1*); 0.5 (*2*); 1.0 (*3*); 2.0 (*4*); 3.0 (*5*); 4.0 (*6*); 5.0 (*7*); 6.0 (*8*); 7.0 (*9*); 8.0 (*10*); 9.0 (*11*); 9.5 (*12*). Symbols mark the experimental data, and the solid straight lines show the linear function approximation. The bold dashed line shows $\ln P_s$, where P_s is the saturated vapor pressure; the dashed lines show the critical pressure and temperature of methane.

As follows from Figure 5a,b, the isosteres of methane adsorption in AC-4 and AC-6 plotted in $\ln P = f(1/T)$ coordinates are well approximated by linear functions. No deviation from the straight lines was observed when passing through the critical temperature of $T_{cr} = 190.77$ K to the nonideality of a gaseous phase. The linearity of adsorption isosteres in the region, where gases show a significant deviation from ideality, is indicative of a specific state of a highly dispersed substance in micropores. This state enables the accumulation of methane in micropores without undergoing any phase transition over wide intervals of sub- and supercritical temperatures and pressures [62].

3.3. Calculation of Thermodynamic Functions of Adsorption Systems

3.3.1. Differential Molar Isosteric Heat of Adsorption

The differential molar isosteric heat of adsorption, q_{st}, is an important thermodynamic parameter, which describes the heat effects of adsorption processes [26,27,30,32–34]. By definition [30], the value of q_{st} is determined as the difference between the molar enthalpy of equilibrium gaseous phase h_g and differential molar enthalpy of adsorption system H_1:

$$q_{st} = h_g - H_1 \tag{4}$$

As noted above, the study was intended to consider the effects associated with the nonideality of a gaseous phase and non-inertness of an adsorbent in terms of the approach by Bakaev [26]. Therefore, we calculated the differential molar isosteric heat of adsorption using an equation:

$$q_{st} = -R \cdot Z \cdot \left[\frac{\partial(\ln P)}{\partial(1/T)} \right]_a \cdot \left[1 - \left(\frac{\partial V_a}{\partial a} \right)_T / v_g \right] - \left(\frac{\partial P}{\partial a} \right)_T \cdot \left[V_a - T \cdot \left(\frac{\partial V_a}{\partial T} \right)_a \right] \tag{5}$$

where $Z = P \cdot v_g/(RT)$ is the coefficient of compressibility of the equilibrium gas phase at pressure P (Pa) and temperature T (K); v_g is the specific gas phase volume, m^3/kg; R is the universal gas constant,

J/(mol·K); $V_a = V_0/m_0$ is the reduced volume of the adsorbent-adsorbate system, cm^3/g; and V_0 and m_0 are the volume and mass of the regenerated adsorbent, respectively. Thus, the Bakaev Equation (4) most fully takes into account the factors which affect the value of differential molar isosteric heat of adsorption: adsorption isothermal deformation $(\partial V_a/\partial a)_T$, temperature isosteric deformation $(\partial V_a/\partial T)_a$, the slopes of the isotherm of adsorption $(\partial P/\partial a)_T$ and isosteres $[\partial \ln P/\partial(1/T)]_a$, and the nonideality of a gas phase Z [26,27].

It was shown in [33] that in the conditions under consideration, the corrections for adsorption deformation of carbon adsorbents upon methane adsorption is minimal (~2–3%) and can be ignored in calculating q_{st}. The data reported by Novikova [63] enabled us to evaluate the maximal value of isosteric temperature deformation $(\partial V_a/\partial T)_a$ and show that the term $T \cdot (\partial V_a/\partial T)_a$ is much lower than V_a in the studied temperature and pressure range. Therefore, we used Equation (5) without corrections for the adsorption-stimulated and thermal deformations of adsorbent:

$$q_{st} = -R \cdot Z \cdot \left[\frac{\partial(\ln P)}{\partial(1/T)}\right]_a - \left(\frac{\partial P}{\partial a}\right)_T \cdot V_a \qquad (6)$$

Figure 6a,b show the functions $q_{st} = f(a)_T$ for AC-4 and AC-6 calculated from the experimental isotherms of methane adsorption following to Equation (6) for different temperatures.

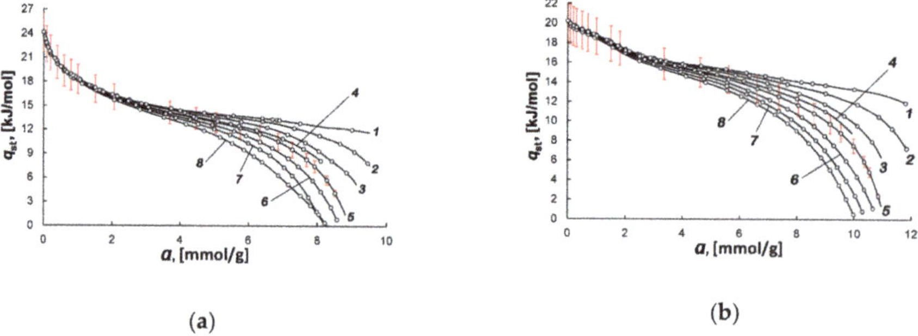

Figure 6. The differential molar isosteric heat of adsorption versus the value of methane adsorption in AC-4 (**a**) and AC-6 (**b**) at temperatures, K: 178 (*1*), 216 (*2*), 243 (*3*), 273.15 (*4*), 300 (*5*), 320 (*6*), 340 (*7*), 360 (*8*). Symbols show experimental data; solid curves are the results of approximation by Equation (6). The error bar is 10%.

The dependences of $q_{st}(a)$ shown in Figure 6a,b are determined by the changes in the methane adsorption mechanisms in ACs as micropores become filled. Indeed, as can be seen from Figure 4a,b, at low pressures (up to 1 Pa), the carbon adsorbents tend to strongly adsorb methane molecules, which occupy a large portion of micropores by binding to high-energy adsorption sites. The high-energy micropores with a width close to the adsorbate molecule diameter with very high adsorbent fields and surface functional groups serve as such adsorption sites. For example, the snapshots and density profiles obtained upon the Grand Canonical Monte Carlo (GCMC) and Molecular Dynamics (MD) simulations of adsorption and dynamic behaviors of methane in the interior of constricted slit models of carbon adsorbents [64] provided evidence that at low pressures methane was adsorbed in the slits with a constriction width close to the molecular diameter, where one layer of adsorbed methane was formed. At low adsorbate loadings, the absolute value of q_{st} depends on the density of adsorption sites, which are the high-energy micropores and heteroatoms, including metal ions. The latter is associated with electrostatic interactions, which are likely to be of minor importance for methane, which has no dipole or quadrupole moments. In any case, the summary effect of these factors determines the lower value of q_{st} found for AC-6 (~20 kJ/mol) compared to AC-4 (~24 kJ/mol). AC-4 differed by a smaller percentage of narrow micropores (see Table 1) and lower content of oxygen than AC-6. Unlike

AC-6, it contains metal ions (see Table 2). According to atomistic simulations [65], a steep initial decrease in q_{st} with a observed in AC-4 is conventionally attributed to heterogeneity in the adsorbent. More precisely, it occurs when the pore size distribution is skewed towards the large pore width, as in the case of AC-4. The reduced slope of $q_{st} = f(a)$ observed at higher values of micropore loading for both adsorbents is likely a result of overlapping the impacts from the adsorbate–adsorbent and adsorbate–adsorbate interactions [65]. When the adsorbate–adsorbate attraction tends to dominate upon the process of methane accumulation, they lead to the formation of adsorption associates of methane molecules in micropores [62,66]. As a result, the heats of methane adsorption in AC-4 and AC-6 come closer together.

As follows from Figure 6a,b, the isosteric heat of adsorption is temperature-independent in the early stages of adsorption. However, with the increase in the value of adsorption, the curves $q_{st} = f(a)_T$ fall as the high-energy adsorption sites become occupied, and the rate of fall of the curves depends on temperature. Therefore, one can observe a so-called "fan" of the curves, each of which correspond to a particular temperature (see Figure 6a,b). With a further increase in pressure, the curves describing the dependency of heats of adsorption diverge according to Equation (5) due to the variations in the temperature-dependent contributions from the compressibility of the gas phase Z and derivative $(\partial P/\partial a) \cdot V_a$ related to the steepness of the adsorption isotherms in the a-P coordinates.

3.3.2. Integral Heat of Methane Adsorption in the Activated Carbons

The methane adsorption/desorption processes are accompanied by the release/absorption of a large amount of heat, which degrades the storage capacity of ANG systems. These thermal effects are estimated by the values of integral heat of adsorption Q:

$$Q(T) = \int_0^a q_{st}(a)_{T=const} da \tag{7}$$

The integral heats of methane adsorption in AC-6 and AC-4 were calculated from the obtained data on isosteric heats of adsorption. The values of Q were used to evaluate the value of ΔT, which is a variation in temperature upon the heating-up/cooling of a 50 L model ANG system filled with AC-4 or AC-6 under adiabatic conditions. In the calculations, we used the specific heat capacity of activated carbons of 0.84 kJ/(kg·K) [67]. Table 3 summarizes the results of the calculations.

Table 3. Thermodynamic characteristics of the adsorbed natural gas (ANG) systems based on AC-4 and AC-6.

AC	T, K	178.00	216.00	243.00	273.15	300.00	320.00	340.00	360.00
	P, MPa	1.0	3.4	6.2	5.1	13.0	16.0	18.0	23.0
AC-4	a, mmol/g	9.6	9.6	9.2	8.1	9.0	8.7	8.3	8.3
	Q, kJ/kg	141.4	135.8	126.6	115.9	116.9	110.4	104.1	99.3
	ΔT, K	58.8	51.9	45.2	46.0	39.4	37.0	35.2	32.6
	P, MPa	1.0	4.0	5.7	5.5	15.0	19.0	21.0	23.0
AC-6	a, mmol/g	11.9	11.9	11.0	10.0	11.1	10.9	10.4	10.1
	Q, kJ/kg	182.1	174.1	159.4	147.7	149.3	140.9	133.8	128.1
	ΔT, K	69.3	58.6	53.8	53.6	44.4	41.4	40.0	38.7

According to Table 3, the increase in temperature produced a decrease in the values of Q. The methane adsorption process in the ANG system based on AC-6, which is the more active adsorbent for methane, is accompanied by the release of a more significant amount of heat compared to that for AC-4. Additionally, the data in Table 3 testify that the heating-up of the ANG system is maximal under adiabatic conditions and low temperatures. At 178 K, ΔT achieved 69 and 59 K for the ANG systems based on AC-6 and AC-4, respectively. With increasing temperature, the heating-up reduces: at 360 K, ΔT is 39 and 33 K for the ANG systems based on AC-6 and AC-4, respectively.

3.3.3. Differential Molar Isosteric Entropy of the Methane-AC Adsorption Systems

The entropy of adsorption system S_1 is an essential thermodynamic state function, which gives insight into the state of adsorbate molecules in micropores, including their interactions in adsorption associates and with the adsorbent. The isosteric molar entropy of an adsorption system S_1 can be calculated using a formula derived from Equation (4) relative to the equilibrium gas phase:

$$S_1 = s_g - \frac{q_{st}}{T} \tag{8}$$

Figure 7a,b show the calculated values of differential molar entropy of the adsorption systems based on AC-4 and AC-6.

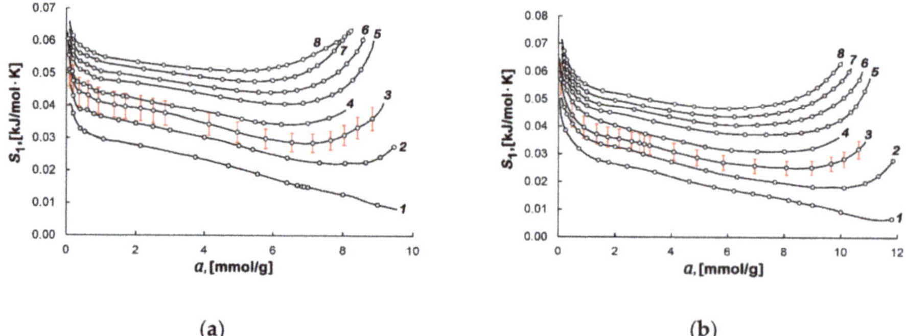

(a)   (b)

Figure 7. The differential molar entropy of the methane-AC-4 (**a**) and methane-AC-6 (**b**) adsorption systems versus the values of methane adsorption at temperatures, K: 178.00 (*1*), 216.00 (*2*), 243.00 (*3*), 273.15 (*4*), 300.00 (*5*), 320.00 (*6*), 340.00 (*7*), and 360.00 (*8*). Symbols indicate the experimental data; solid lines are the smoothing spline curves. The error bar is 10%.

Figure 7a,b demonstrate a steep initial decrease in the entropy at $a < 0.5$ mmol/g, which is attributed to the occupation of the most strongly adsorbing sites. In the region of intermediate values of adsorption, from 1 to 7 mmol/g, the decrease in the entropy slows down due to further loading of micropores and the formation of the adsorption associates caused by increasing mutual attraction between adsorbate molecules [66]. When micropore fillings are high, a sharp rise in curves $S_1(a)$ is observed, which can be interpreted as a result of the rearrangement of the adsorbate structure, leading to the formation of the denser methane associates. As was shown in [32,33], a further increase in methane adsorption leads to a weak local peak followed by a drop in the entropy at high fillings of micropores. These variations in the entropy are indicative of the completion of the formation of adsorption associates and subsequent rearrangement of adsorbed molecules caused by the increased repulsion between them.

3.3.4. Differential Molar Isosteric Enthalpy of the Methane-AC Adsorption Systems

The enthalpy of an adsorption system, H_1, is a thermodynamic state function, which determines an amount of energy, which the system can potentially transform into heat—in other words, the heat content:

$$H_1 = h_g - q_{st} \tag{9}$$

Figure 8a,b shows the results of the calculation of H_1 for the studied adsorption systems at different temperatures.

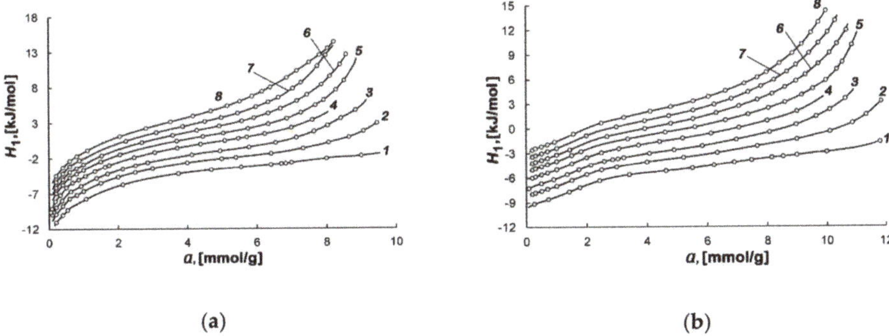

(a)

(b)

Figure 8. Dependence of the differential molar isosteric enthalpy of the methane-AC-4 (**a**) and methane-AC-6 (**b**) adsorption systems on the value of adsorption at temperatures, K: 178.00 (*1*), 216.00 (*2*), 243.00 (*3*), 273.15 (*4*), 300.00 (*5*), 320.00 (*6*), 340.00 (*7*), and 360.00 (*8*). Symbols indicate the experimental data; solid lines are the smoothing spline curves.

As follows from Figure 8a,b, the enthalpy is dependent on temperature, and this dependence becomes stronger with the increase in the micropore fillings. The negative values of enthalpy H_1 are determined by the reference level accepted for the standard state of the gas phase—h_g.

Figure 9a,b shows the temperature dependences of differential molar enthalpy of the methane-AC-4 and methane-AC-6 systems at various values of adsorption.

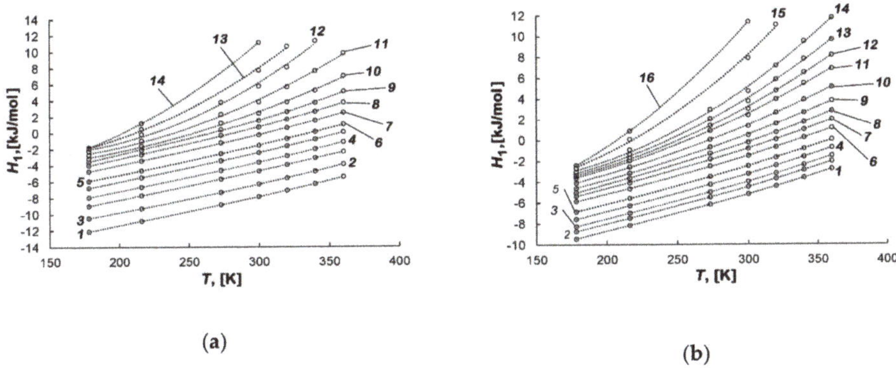

(a)

(b)

Figure 9. Temperature dependence of the differential molar enthalpy of the methane-AC-4 (**a**) and methane-AC-6 (**b**) adsorption systems at the values of adsorption *a*, mmol/g: 0.1 (*1*), 0.3 (*2*), 0.6 (*3*), 1 (*4*), 1.5 (*5*), 2 (*6*), 3 (*7*), 4 (*8*),5 (*9*), 6 (*10*), 7 (*11*), 7.8 (*12*), 8.3 (*13*), 8.8 (*14*) (**a**) and 0.1 (*1*), 0.6 (*2*), 1 (*3*), 1.5 (*4*), 2 (*5*), 3 (*6*), 4 (*7*), 5 (*8*), 6 (*9*), 7 (*10*), 8 (*11*), 8.5 (*12*), 9 (*13*), 9.5 (*14*), 10.5 (*15*), 11 (*16*) (**b**).

Figure 9a,b shows that, at low loading, the enthalpy of these systems increases almost linearly, which is caused by the temperature invariance $q_{st} \neq f(T)$ and the linear dependence of h_g on temperature. However, with the increase in the micropore fillings, the dependences $H_1(T)$ become nonlinear. The rate of their growth, $H_1'(T)$, increases with temperature, which is determined by the contribution from temperature-dependent compressibility of the gas phase Z and derivative $(\partial P/\partial a)\cdot V_a$ in Equation (6) for q_{st} (see the fan of curves $q_{st} = f(a)$ in Figure 6a,b).

The variations in H_1 with *a* and *T* obtained from the experimental data for both methane-AC adsorption systems are consistent with those calculated for model micropores in many numerical experiments. As was found in [68] by using the continuous fractional component Monte Carlo algorithm, the growth of enthalpy at high micropore loadings close to saturation is caused by repulsive interactions arising between an introduced adsorbate molecule and both the previously adsorbed species and the adsorbent. The appearance of these intermolecular repulsive forces leads to the rearrangement of the adsorption associates.

3.3.5. Differential Molar Isosteric Heat Capacity of the Methane-AC Adsorption Systems

In contrast to the specific heat capacity of the bulk phase, the specific heat capacity of any adsorption system is dependent not only on pressure, temperature but on the value of adsorption. Therefore, a comprehensive thermodynamic description of an adsorption system includes the differential molar isosteric heat of adsorption—C_a. The values of C_a were calculated using the Kirchhoff equation derived by the differentiation of Equation (4) over the temperature:

$$C_a = \left(\frac{\partial H_1}{\partial T}\right)_a = \left(\frac{\partial h_g}{\partial T}\right)_a - \left(\frac{\partial q_{st}}{\partial T}\right)_a \tag{10}$$

Figure 10a,b shows the temperature dependencies of the differential molar isosteric heat capacities of the methane-AC-4 and methane-AC-6 adsorption systems, respectively.

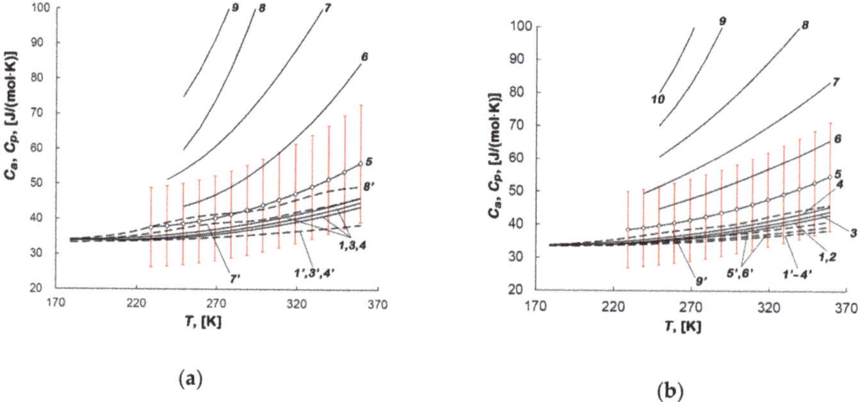

(a)   (b)

Figure 10. Temperature dependences of the differential molar isosteric heat capacities of the methane-AC-4 (**a**) and methane-AC-6 (**b**) adsorption systems and the gas phase (numbering with stroke) at the values of methane adsorption, mmol/g: 0.1 (*1*), 0.3 (*2*), 1.0 (*3*), 2.0 (*4*), 4.0 (*5*), 6.0 (*6*), 7.0 (*7*), 7.8 (*8*), 8.3 (*9*). The error bar is 30%.

Figure 10a,b shows that the heat capacity of the adsorption systems is identical in magnitude to the isobaric heat capacity of the gas system (see 1, 2, 3, 4 and 1′, 2′, 4′ curves). However, when the temperatures are above 220 K, the differences become more significant, and the heat capacity of the adsorption system exceeds the values of C_p. Furthermore, with the increase in the methane adsorption values and temperature, the heat capacity also increases. Moreover, the $C_a(T)$ curves diverge more strongly as the temperature increases.

As follows from Figure 10a,b, the isosteric heat capacities of the methane-AC-4 and methane -AC-6 adsorption systems differ significantly from each other at the same adsorption values. Indeed, at $a \sim 8$ mmol/g and 270 K, the isosteric heat capacity of the methane-AC-4 system is 45% more than that in the adsorption system with AC-6. Therefore, one can conclude that the isosteric heat capacity is lower when methane is adsorbed in AC-6 with a higher volume of micropores and less content of carbon species compared to AC-4.

According to [32], which summarized the thermodynamic functions of a large number of gas/microporous material adsorption systems, in most cases, the $C_a(T)$ dependence exhibited a local maximum at elevated temperatures. When adsorption achieved the high values, this maximum shifted towards lower temperatures, as was the cases of Xe [69] and CH_4 adsorption [70,71] in NaX zeolites, and CH_4 adsorption in the microporous silicon carbide-derived activated carbon labeled as AUK [33]. Therefore, we suggest that these regularities are general and will be observed in the studied adsorption systems with the increase in the values of methane adsorption, pressure, and temperature. Such features of the $Ca(T)$ dependences cannot be caused by the gaseous phase, since these dependencies significantly diverge; for example, curves 7 and 7'. The adsorption-induced deformation of activated carbon, which is the component of the adsorption system, is also not responsible for this effect, since the changes in its crystal structure are imperceptible. The only logical explanation for this effect is a change in the state of adsorbed molecules, namely the formation of adsorption associates in micropores, which is confirmed by the results of molecular dynamics simulations [34,62,66].

4. Conclusions

The activated carbons were synthesized from peat of high degree of metamorphism in three stages: carbonization, thermochemical K_2S activation at 1053–1133 K, and steam activation at 1173 K. The stage of steam activation performed for two different periods resulted in two different AC samples: AC-4 and AC-6. The data of SEM, X-ray, and adsorption investigations revealed that the most activated carbon possessed the largest micropore volume and specific BET surface. The increase in the activation time caused the changes in the chemical composition and crystallinity. A combination of these changes determines a higher adsorption capacity of AC-6 with respect to methane. The results indicate that a detailed comparative analysis of the adsorption performance, characteristics of porous structure (SEC), and data on phase and chemical composition for activated carbons prepared from various precursors under different activation conditions is needed to establish a precise correlation between textural and adsorption behaviors. This correlation is essential for determining a criterium for an optimal adsorbent for ANG storage.

The general thermodynamic equation by Bakaev made it possible to calculate the differential molar isosteric heats of methane adsorption in AC-4 and AC-6 at pressures up to 25 MPa in the temperature range from 178 to 360 K. The differential molar isosteric entropy, enthalpy, and heat capacity evaluated as the functions of temperature and methane adsorption for both methane-AC adsorption systems are affected by heterogeneous porous, phase and chemical structure of the adsorbents. It should be noted that despite the differences in absolute values, the variations in the thermodynamic functions with temperature and adsorption values reflect the general pattern of changes in the state of adsorbed molecules upon the adsorption process in the heterogeneous microporous carbons: from binding with the high-energy adsorption sites to the formation of the adsorption molecular associates, and their subsequent rearrangement close to saturation.

The heating-up of both the ANG systems calculated from the integrated heat of adsorption was found to be maximal under adiabatic conditions and low temperatures. The magnitudes of the temperature changes of the ANG system based on AC-4 upon the adsorption (charging) process are lower than those found for that with AC-6. When the temperature of the adsorption process grows, the heating-up of both adsorption systems decreases.

The isosteric heat capacity of the methane adsorption system based on AC-4 with a lower degree of activation and higher energy of adsorption exceeded that for the system with AC-6. When the value of adsorption is close to 8 mmol/g, and the temperature is 270 K, the difference between the values of C_a for the systems based on AC-4 and AC-6 amounts to ~ 45%.

Finally, the evaluation of the key thermodynamic quantities is essential for selecting an efficient adsorbent, designing both the ANG systems and heat-exchange facilities to achieve the maximum of both storage density and deliverable capacity.

Author Contributions: Conceptualization, I.M. and A.F.; methodology, A.F., A.S.; software, I.M., A.S.; validation, A.F., A.S.; formal analysis, I.M.; investigation, I.M., A.S.; resources, A.S.; data curation, A.F., E.K.; writing—original draft preparation, I.M., E.K.; writing—review and editing, E.K.; visualization, E.K.; supervision, A.F.; project administration, A.F., I.M.; funding acquisition, A.F. All authors have read and agreed to the published version of the manuscript.

Funding: The research was carried out within the State Assignment of the Russian Federation (Project No. 01201353185) and the plan of the RAS Scientific Council (Theme No. 20-03-460-01).

Acknowledgments: The investigations were carried out with the use of equipment of the Center of Physical Methods of Investigations of the A.N. Frumkin Institute of Physical Chemistry and Electrochemistry of Russian Academy of Sciences. We thank A.A. Shiryaev and V.V. Vysotskii for help in the XRD, SAXS, and SEM experiments and constructive suggestions, and V.A. Bakaev for critical reading and participating in discussions.

Conflicts of Interest: The authors declare no conflict of interest. The funders had no role in the design of the study; in the collection, analyses, or interpretation of data; in the writing of the manuscript, or in the decision to publish the results.

References

1. Nie, Z.; Lin, Y.; Jin, X. Research on the theory and application of adsorbed natural gas used in new energy vehicles: A review. *Front. Mech. Eng.* **2016**, *11*, 258–274. [CrossRef]
2. Tsivadze, A.Y.; Aksyutin, O.E.; Ishkov, A.G.; Men'shchikov, I.E.; Fomkin, A.A.; Shkolin, A.V.; Khozina, E.V.; Grachev, V.A. Porous carbon-based adsorption systems for natural gas (methane) storage. *Russ. Chem. Rev.* **2018**, *87*, 950–983. [CrossRef]
3. Kumar, K.V.; Preuss, K.; Titirici, M.M.; Rodríguez-Reinoso, F. Nanoporous materials for the onboard storage of natural gas. *Chem. Rev.* **2017**, *117*, 1796–1825. [CrossRef]
4. Men'shchikov, I.E.; Fomkin, A.A.; Tsivadze, A.Y.; Shkolin, A.V.; Strizhenov, E.M.; Khozina, E.V. Adsorption accumulation of natural gas based on microporous carbon adsorbents of different origin. *Adsorption J.* **2017**, *23*, 327–339. [CrossRef]
5. Elizabeth Casco, M.; Martínez-Escandell, M.; Gadea-Ramos, E. High-pressure methane storage in porous materials: Are carbon materials in the pole position? *Chem. Mater.* **2015**, *27*, 959–964. [CrossRef]
6. Fomkin, A.A.; Men'shchikov, I.E.; Pribylov, A.A.; Shkolin, A.V.; Zaitsev, D.S.; Tvardovski, A.V. Methane adsorption in microporous carbon adsorbent with wide distribution of pores. *Colloid J.* **2017**, *79*, 144–151. [CrossRef]
7. Shevchenko, A.O.; Pribylov, A.A.; Zhedulov, S.A.; Men'shchikov, I.E.; Shkolin, A.V.; Fomkin, A.A. Methane adsorption in microporous carbon adsorbent LCN obtained by thermochemical synthesis from lignocellulose. *Prot. Met. Phys. Chem. Surf.* **2019**, *55*, 211–216. [CrossRef]
8. Policicchio, A.; Maccallini, E.; Agostino, R.G.; Ciuchi, F.; Aloise, A.; Giordano, G. Higher methane storage at low pressure and room temperature in new easily scalable large-scale production activated carbon for static and vehicular applications. *Fuel* **2013**, *104*, 813–821. [CrossRef]
9. Tsivadze, A.Y.; Aksyutin, O.E.; Ishkov, A.G.; Knyazeva, M.K.; Solovtsova, O.V.; Men'shchikov, I.E.; Fomkin, A.A.; Shkolin, A.V.; Khozina, E.V.; Grachev, V.A. Metal-organic framework structures: Adsorbents for natural gas storage. *Russ. Chem. Rev.* **2019**, *88*, 925–978. [CrossRef]
10. Makal, T.A.; Li, J.-R.; Lu, W.; Zhou, H.-C. Methane storage in advanced porous materials. *Chem. Soc. Rev.* **2012**, *41*, 7761–7779. [CrossRef]
11. Yuan, D.; Lu, W.; Zhao, D.; Zhou, H.-C. Highly stable porous polymer networks with exceptionally high gas-uptake capacities. *Adv. Mater.* **2011**, *23*, 3723–3725. [CrossRef]
12. Rubio-Martinez, M.; Avci-Camur, C.; Thornton, A.W.; Imaz, I.; Maspoch, D.; Hill, M.R. New synthetic routes towards MOF production at scale. *Chem. Soc. Rev.* **2017**, *46*, 3453–3480. [CrossRef]
13. Valizadeh, B.; Nguyen, T.N.; Stylianou, K.C. Shape engineering of metal-organic frameworks. *Polyhedron* **2018**, *145*, 1–15. [CrossRef]
14. Chapman, K.W.; Halder, G.J.; Chupas, P.J. Pressure-induced amorphization, and porosity modification in a metal-organic framework. *J. Am. Chem.* **2009**, *131*, 17546–17547. [CrossRef] [PubMed]
15. Tagliabue, M.; Rizzo, C.; Millini, R.; Dietzel, P.D.C.; Blom, R.; Zanardi, S. Methane storage on CPO-27-Ni pellets. *J. Porous Mater.* **2011**, *18*, 289–296. [CrossRef]
16. Zhang, H.; Deria, P.; Farha, O.K.; Hupp, J.T.; Snurr, R.Q. A thermodynamic tank model for studying the effect of higher hydrocarbons on natural gas storage in metal-organic frameworks. *Energy Environ. Sci.* **2015**, *8*, 1501–1510. [CrossRef]

17. DeSantis, D.; Mason, J.A.; James, B.D.; Houchins, C.; Long, J.R.; Veenstra, M. Techno-economic analysis of metal−organic frameworks for hydrogen and natural gas storage. *Energy Fuels* **2017**, *31*, 2024–2032. [CrossRef]

18. Rodríguez-Reinoso, F. Activated carbon and adsorption. In *Encyclopedia of Materials: Science and Technology*; Elsevier: Amsterdam, The Netherlands, 2001; pp. 22–34.

19. Uraki, Y.; Tamai, Y.; Ogawa, M.; Gaman, S.; Tokurad, S. Preparation of activated carbon from peat. *BioResources* **2009**, *4*, 205–213.

20. Mukhin, V.M.; Tarasov, A.V.; Klushin, V.N. *Aktivnie Ugli Rossii (Active Carbons of Russia)*; Metallurgiya: Moscow, Russia, 2000; p. 352.

21. Dubinin, M.M. Physical adsorption of gases and vapors in micropores. *Prog. Surf. Membr. Sci.* **1975**, *9*, 1–70.

22. WanDaud, W.M.A.; Houshamnd, A.H. Textural characteristics, surface chemistry, and oxidation of activated carbon. *J. Nat. Gas. Chem.* **2010**, *19*, 267–279.

23. Myers, A.L. Thermodynamics of adsorption in porous materials. *AIChE J.* **2002**, *48*, 145–160. [CrossRef]

24. Wang, Y.; Ercan, C.; Khawajah, A.; Othman, R. Experimental and theoretical study of methane adsorption on granular activated carbons. *AIChE J.* **2012**, *58*, 782–788. [CrossRef]

25. Rahman, K.A.; Loh, W.S.; Ng, K.C. Heat of adsorption and adsorbed phase specific heat capacity of methane/activated carbon system. *Procedia Eng.* **2013**, *56*, 118–125. [CrossRef]

26. Bakaev, V.A. Molecular Theory of Physical Adsorption. Ph.D. Thesis, Moscow State University, Moscow, Russia, May 1990. (In Russian)

27. Bakaev, V.A. One possible formulation of the thermodynamics of sorption equilibrium. *Bull. Acad. Sci. USSR Div. Chem. Sci.* **1971**, *20*, 2516–2520. [CrossRef]

28. Guggenheim, E.A. *Modern Thermodynamics by the Methods of Willard Gibbs*; Methuen & Co. Ltd.: London, UK, 1933; p. 206.

29. Guggenheim, E.A. *Thermodynamics. An Advanced Treatment for Chemist and Physicist*; North Holland Publishing Company: Amsterdam, The Netherlands, 1967; pp. 166–169.

30. Hill, T.L. *Theory of Physical Adsorption in Advances in Catalysis and Related Subjects*; Academic Press: New York, NY, USA, 1952; Volume 4, pp. 211–258.

31. Gibbs, J.V. *Thermodynamic Works*; Gostekhizdat: Moscow, Russia, 1950; p. 492. (In Russian)

32. Fomkin, A.A. Adsorption of gases, vapors and liquids by microporous adsorbents. *Adsorption* **2005**, *11*, 425–436. [CrossRef]

33. Shkolin, A.V.; Fomkin, A.A. Thermodynamics of methane adsorption on the microporous carbon adsorbent ACC. *Russ. Chem. Bull.* **2008**, *57*, 1799–1805. [CrossRef]

34. Strizhenov, E.M.; Shkolin, A.V.; Fomkin, A.A.; Sinitsyn, V.A.; Zherdev, A.A.; Smirnov, I.A.; Pulin, A.L. Low-temperature adsorption of methane on microporous AU-1 carbon adsorbent. *Prot. Met. Phys. Chem. Surf.* **2014**, *50*, 15–21.

35. Jinmung, H.; Xuehui, M. Physical and chemical properties of peat. In *Coal, Oil Shale, Natural Bitumen, Heavy Oil and Peat*; EOLSS Publisher Co Ltd.: Oxford, UK, 2009; Volume II, pp. 315–318.

36. Fenelonov, V.B. *Poristyi Uglerod (Porous Carbon)*; Boreskov Institute of Catalysis SB RAS: Novosibirsk, Russia, 1995; p. 518. (In Russian)

37. Manocha, S.M. Porous carbons. In *Frontiers in Material Science*; Universities Press Private Ltd.: Hyderabad, India, 2005; pp. 335–348.

38. Rodriguez-Reinoso, F.; Molina-Sabio, M.; Gonzalez, M.T. The use of steam and CO_2 as activating agents in the preparation of activated carbons. *Carbon* **1995**, *33*, 15–23. [CrossRef]

39. Molina-Sabio, M.; González, M.T.; Rodriguez-Reinoso, F.; Sepúlveda-Escribano, A. Effect of steam and carbon dioxide activation in the micropore size distribution of activated carbon. *Carbon* **1996**, *34*, 505–509. [CrossRef]

40. Sychev, V.V.; Vasserman, A.A.; Zagoruchenko, V.A.; Kozlov, A.D.; Spiridonov, G.A.; Tzymarnyi, V.A. *Thermodynamic Properties of Methane*; Izd. Standartov: Moscow, Russia, 1979; p. 349. (In Russian)

41. Shkolin, A.V.; Fomkin, A.A. Measurement of carbon-nanotube adsorption of energy-carrier gases for alternative energy systems. *Meas. Techn.* **2018**, *61*, 395–401. [CrossRef]

42. Shkolin, A.V.; Fomkin, A.A.; Men'shchikov, I.E.; Kharitonov, V.M.; Pulin, A.L. A bench for measuring the adsorption of gases and vapors by gravimetric technique and the method of its operational use. RF Patent 2019143065, 23 December 2019.

43. GOST 34100.3-2017/ISO/IEC Guide 98-3:2008. Part 3. Uncertainty of measurement. Part 3. Guide to the expression of uncertainty in measurement. Available online: https://files.stroyinf.ru/Data/651/65118.pdf (accessed on 18 June 2020).

44. Brunauer, S.; Emmett, P.H.; Teller, E. Adsorption of gases in multimolecular layers. *J. Am. Chem. Soc.* **1938**, *60*, 309–319. [CrossRef]

45. Isirikyan, A.A.; Kiselev, A.V. The absolute adsorption isotherms of vapors of nitrogen, benzene, and n-hexane, and the heats of adsorption of benzene and n-hexane on graphitized carbon black. *J. Phys. Chem.* **1961**, *65*, 601–607. [CrossRef]

46. Kiselev, A.V. New adsorption method for the determination of surface of adsorbent. *Uspekhi Khimii* **1945**, *XVI*, 367–394. (In Russian)

47. Shkolin, A.V.; Fomkin, A.A.; Pulin, A.L.; Yakovlev, V.Y. A technique for measuring an adsorption-induced deformation. *Instrum. Exp. Tech.* **2008**, *51*, 150–155. [CrossRef]

48. Fomkin, A.A.; Shkolin, A.V.; Men'shchikov, I.E.; Pulin, A.L.; Pribylov, A.A.; Smirnov, I.A. Measurement of adsorption of methane at high pressures for alternative energy systems. *J. Meas. Techn.* **2016**, *58*, 1387–1391. [CrossRef]

49. Pribylov, A.A.; Serpinskii, V.V.; Kalashnikov, S.M. Adsorption of gases by microporous adsorbents under pressures up to hundreds of megapascals. *Zeolites* **1991**, *11*, 846–849. [CrossRef]

50. Downarowicz, D. Adsorption characteristics of propane-2-ol vapours on activated carbon Sorbonorit 4 in electrothermal temperature swing adsorption process. *Adsorption* **2015**, *21*, 87–98. [CrossRef]

51. Bergna, D.; Hu, T.; Prokkola, H.; Romar, H.; Lassi, U. Effect of some process parameters on the main properties of activated carbon produced from peat in a lab-scale process. *Waste Biomass Valorization* **2020**, *11*, 2837–2848. [CrossRef]

52. Zhu, Y.; Murali, S.; Stoller, M.D.; Ganesh, K.J.; Cai, W.; Ferreira, P.J.; Pirkle, A.; Wallace, R.M.; Cychosz, K.A.; Thommes, M.; et al. Carbon-based supercapacitors produced by activation of graphene. *Science* **2011**, *332*, 1537–1541. [CrossRef] [PubMed]

53. Feigin, L.A.; Svergun, D.I. *Structure Analysis by Small-Angle X-Ray and Neutron Scattering*; Plenum Press: New York, NY, USA, 1989; p. 335.

54. Shiryaev, A.A.; Voloshchuk, A.M.; Volkov, V.V.; Averin, A.A.; Artamonova, S.D. Nanoporous active carbons at ambient conditions: A comparative study using X-ray scattering and diffraction, Raman spectroscopy and N_2 adsorption. *J. Phys. Conf. Series* **2017**, *848*, 012009. [CrossRef]

55. Guinier, A. La diffraction des rayons X aux tres petits angles: Application a l'etude de phenomenes ultramicroscopiques. *Ann. Phys.* **1939**, *11*, 161–237. [CrossRef]

56. Dubinin, M.M.; Plavnik, G.M. Microporous structures of carbonaceous adsorbents. *Carbon* **1968**, *6*, 183–192. [CrossRef]

57. Dubinin, M.M.; Erashko, I.T.; Kadlec, O.; Ulin, V.I.; Voloshchuk, A.M.; Zolotarev, P.P. Kinetics of physical adsorption by carbonaceous adsorbents of biporous structure. *Carbon* **1975**, *13*, 193–200. [CrossRef]

58. Dubinin, M.M.; Stoeckli, H.F. Homogeneous and heterogeneous micropore structures in carbonaceous adsorbents. *J. Colloid Interface Sci.* **1980**, *75*, 34–42. [CrossRef]

59. Bakaev, V.A. The statistical thermodynamics of adsorption equilibriums in the case of zeolites. *Dokl. Acad. Nauk SSSR* **1966**, *167*, 369–372.

60. Pribylov, A.A.; Murdmaa, K.A. Adsorption of gases onto Polymer Adsorbent MN-270 in the region of supercritical temperatures and pressures. *Prot. Met. Phys. Chem. Surf.* **2020**, *56*, 115–121. [CrossRef]

61. Pribylov, A.A.; Kalinnikova, I.A.; Murdmaa, K.O. Determination of the average heat and characteristic energy from the adsorption isotherm. *Russ. Chem. Bull. Int. Ed.* **2016**, *65*, 972–977. [CrossRef]

62. Shkolin, A.V.; Fomkin, A.A.; Tsivadze, A.Y.; Anuchin, K.M.; Men'shchikov, I.E.; Pulin, A.L. Experimental study and numerical modeling: Methane adsorption in microporous carbon adsorbent over the subcritical and supercritical temperature regions. *Prot. Met. Phys. Chem. Phys. Surf.* **2016**, *52*, 955–963. [CrossRef]

63. Novikova, S.I. *Teplovoe Rasshirenie Tverdykh Tel (Heat Expansion of Solids)*; Nauka: Moscow, Russia, 1974; p. 293. (In Russian)

64. Yeganegi, S.; Gholampour, F. Methane adsorption and diffusion in a model nanoporous carbon: An atomistic simulation study. *Adsorption* **2013**, *19*, 979–987. [CrossRef]

65. Nicholson, D.; Stubos, T. Simulation of adsorption in micropores. In *Recent Advances in Gas Separation by Microporous Ceramic Membranes*; Elsevier Science B.V.: Amsterdam, The Netherlands, 2000; Volume 6, pp. 231–256.

66. Anuchin, K.M.; Fomkin, A.A.; Korotych, A.P.; Tolmachev, A.M. Adsorption concentration of methane. dependence of adsorbate density on the width of slit-shaped micropores in activated carbons. *Prot. Met. Phys. Chem. Surf.* **2014**, *50*, 173–177. [CrossRef]

67. Kel'tsev, N.V. *Osnovy Adsorbtsionnoi Praktiki (Fundamentals of Adsorption Practice)*; Khimiya: Moscow, Russia, 1976; p. 90. (In Russian)

68. Torres-Knoop, A.; Poursaeidesfahani, A.; Vlugt, T.J.H.; Dubbeldam, D. Behavior of the enthalpy of adsorption in nanoporous materials close to saturation conditions. *J. Chem. Theor. Comput.* **2017**, *13*, 3326–3339. [CrossRef] [PubMed]

69. Fomkin, A.A.; Serpinskii, V.V.; Bering, B.P. Investigation of the adsorption of xenon on NaX zeolite within a broad range of pressures and temperatures. *Russ. Chem. Bull.* **1975**, *24*, 1147–1150. [CrossRef]

70. Chkhaidze, E.V.; Fomkin, A.A.; Serpinskii, V.V.; Tsitsishvili, G.V.; Dubinin, M.M. Methane adsorption on a microporous carbon adsorbent in the precritical and hypercritical regions. *Russ. Chem. Bull.* **1986**, *35*, 847–849. [CrossRef]

71. Chkhaidze, E.V.; Fomkin, A.A.; Serpinskii, V.V.; Tsitsishvili, G.V. Thermodynamics of the adsorption of methane on NaX zeolite. *Russ. Chem. Bull.* **1986**, *35*, 252–255. [CrossRef]

 *nanomaterials*

Article

Statistical Mechanics at Strong Coupling: A Bridge between Landsberg's Energy Levels and Hill's Nanothermodynamics

Rodrigo de Miguel [1,*] **and J. Miguel Rubí** [2]

[1] Department of Teacher Education, Norwegian University of Science and Technology, 7491 Trondheim, Norway

[2] Department of Condensed Matter Physics, University of Barcelona, 08007 Barcelona, Spain; mrubi@ub.edu

* Correspondence: rodrigo.demiguel@ntnu.no; Tel.: +47-73412115

Received: 21 November 2020; Accepted: 6 December 2020; Published: 10 December 2020

Abstract: We review and show the connection between three different theories proposed for the thermodynamic treatment of systems not obeying the additivity ansatz of classical thermodynamics. In the 1950s, Landsberg proposed that when a system comes into contact with a heat bath, its energy levels are redistributed. Based on this idea, he produced an extended thermostatistical framework that accounts for unknown interactions with the environment. A decade later, Hill devised his celebrated *nanothermodynamics*, where he introduced the concept of *subdivision potential*, a new thermodynamic variable that accounts for the vanishing additivity of increasingly smaller systems. More recently, a thermostatistical framework *at strong coupling* has been formulated to account for the presence of the environment through a Hamiltonian of mean force. We show that this modified Hamiltonian yields a temperature-dependent energy landscape as earlier suggested by Landsberg, and it provides a thermostatistical foundation for the subdivision potential, which is the cornerstone of Hill's nanothermodynamics.

Keywords: thermodynamics at strong coupling; *temperature-dependent* energy levels; thermodynamics of small systems; nanothermodynamics

---

1. Introduction

Systems are always in contact with an environment that influences their energy, volume, and mass. In certain cases, the presence of the environment is of little significance and, for simplicity, the system may be described as though it were isolated. In other cases, surroundings significantly affect the properties of systems, and external interactions need to be taken into account. Systems subject to the latter scenario may be referred to as small, where *small* is not an attribute determined by the system's sheer size, but rather by how the size compares to the range of the interactions affecting the system [1–3].

In contrast to their macroscopic analogues, small systems are *non-extensive*, and consequently, also *non-additive*, i.e., the total system is not just a sum of its parts. Clearly, this lack of additivity escapes the paradigms of classical thermodynamics, and a modified thermophysical theory is needed to describe small systems.

In 1940, Rushbrooke proposed a novel statistical mechanical method to account for the non-additivity of a system whose interactions with the environment are unknown [4]. This method, which consisted on allowing the energy levels of the system to depend on the temperature of the bath, was later refined by Landsberg in the 1950s [5,6]. In the decades that followed, Landsberg's theory of *temperature-dependent* energy levels was directly invoked to study the temperature dependence of energy gaps in semiconductors [7], as well as bosonic systems where the standard thermodynamic description breaks down due to strong interactions [8,9]. Temperature-dependent energy levels have since become ubiquitous in the study of semiconductors (see e.g., [10,11] and references therein) and bosonic systems (see e.g., [12] and references therein). The theory has also been applied to superconductivity [13], optomechanical oscillators [14], irreversible effects in thermoelectric phenomena [15,16], small-system thermalization [17], and even information theory [18].

Despite the many applications of Landsberg's framework, nobody used it to develop an alternative thermodynamic theory for nonadditive systems. In fact, the first thermodynamic theory valid for nonadditive systems did not appear until 1962, when Hill independently proposed his celebrated thermodynamics of small systems [19,20], later termed *nanothermodynamics* [21–23]. In contrast to Landsberg's approach, Hill's thermodynamic theory is not based on modified energy levels, but rather on the introduction of a *subdivision potential* to account for nonadditive effects. Hill's theory has proven fruitful to describe nanoscale phenomena [24–29], and also the nonadditivity that occurs in macroscopic systems subject to long-range interactions [2].

In recent times, Landsberg's idea of a statistical mechanics with temperature-dependent energy levels has covertly reappeared in a framework known as statistical mechanics *at strong coupling* [30]. This framework is based on the formulation of a *Hamiltonian of mean force*, which accounts for environmental perturbations on the small system of interest [30]. This thermostatistical approach has recently been applied to describe nanomechanical systems [31], nanoscale interfacial phenomena [3] and the advent of negative thermophoresis [32].

In this work, we show that Landsberg's physical theory of temperature-dependent energy levels [6] is the foundation of the Hamiltonian of mean force upon which the recent statistical mechanics at strong coupling [30] is based. Furthermore, we show that statistical mechanics at strong coupling provides a thermostatistical foundation for the subdivision potential, which, as we shall see below, is the cornerstone of Hill's nanothermodynamics [20]. Although the inception and evolution of these three powerful theories has been completely independent, and despite their apparent differences, they are indeed indeed deeply connected.

The remainder of this paper is structured as follows. In Section 2, we show that Landsberg's theory of temperature-dependent energy levels lies at the heart of the Hamiltonian of mean force upon which statistical mechanics at strong coupling is based. In Section 3, we show that statistical mechanics at strong coupling provides a thermostatistical basis for the subdivision potential, the thermodynamic variable upon which Hill's nanothermodynamics is based. Concluding remarks are given in Section 4.

2. 'Temperature-Dependent' Energy Levels and the Hamiltonian of Mean Force

In 1940, Rushbrooke proposed a generalized *statistical mechanics for assemblies whose energy-levels depend on the temperature*. His idea was that *the physical state of either part of the assembly will depend on the temperature, and thus so will its interaction with the other part of the assembly* [4]. Rushbrooke's treatment was further generalized by Landsberg in the 1950s, when he applied the method to explore the equilibrium properties of systems subject to unknown interactions with some environment [5,6]. In the following, we show that Landsberg's theory leads to the Hamiltonian of mean force used in statistical mechanics at strong coupling [30].

Landsberg's theory may be summarized as follows. Considering an ensemble of systems at temperature T, one may abstractly decompose each system into two interacting parts: part 1 in state i_1 and with energy $E_1(i_1)$; and part 2 in state i_2 and with energy $E_2(i_2)$. It is then assumed that, while the effective energy of part 2 does not deviate from its isolated value, the interaction between both subsystems does affect the energy of part 1. The total energy of the interacting system may then be written as

$$E(i_1, i_2) = E_1(i_1, i_2) + E_2(i_2), \tag{1}$$

and the effective energy levels $\mathcal{E}_1(i_1)$ of part 1 become (see § 4 in [6])

$$\mathcal{E}_1(i_1) = -k_B T \ln \left[\frac{\sum\limits_{i_2} e^{-E_1(i_1,i_2)/k_B T}\, e^{-E_2(i_2)/k_B T}}{\sum\limits_{i_2} e^{-E_2(i_2)/k_B T}} \right]. \tag{2}$$

This method is useful when there is incomplete statistical mechanical information, and the interactions between the system (part 1) and the environment (part 2) are unknown. The modified energy levels $\mathcal{E}_1(i_1)$ are anomalous in that they are effectively temperature-dependent. Inserting these effective energy levels into the partition function

$$\mathcal{Z}_1 = \sum\limits_{i_1} e^{-\mathcal{E}_1(i_1)/k_B T} \tag{3}$$

results in effective expressions for the internal energy $\mathcal{U}_1$ and the entropy $\mathcal{S}_1$ of the system

$$\mathcal{U}_1 = k_B T^2 \frac{\partial}{\partial T} \log \mathcal{Z}_1 = \langle \mathcal{E}_1(i_1) \rangle - T \left\langle \frac{d\mathcal{E}_1(i_1)}{dT} \right\rangle, \tag{4}$$

$$\mathcal{S}_1 = \frac{\mathcal{U}_1}{T} + k_B \ln \mathcal{Z}_1 = \frac{\langle \mathcal{E}_1(i_1) \rangle}{T} + k_B \ln \mathcal{Z}_1 - \left\langle \frac{d\mathcal{E}_1(i_1)}{dT} \right\rangle, \tag{5}$$

which differ from the usual additive expressions by the incidence of the last terms (see § 2 in [6]).

It should be noted that, in the approximation in which interactions are neglected, then $E_1(i_1, i_2)$ simply becomes $E_1(i_1)$, and the temperature-dependent energy levels $\mathcal{E}_1(i_1)$ reduce to the temperature-independent, purely mechanical energies $E_1(i_1)$. In general, however, temperature-dependence does emerge into effective energy levels. Indeed, when examining energy levels in polarons, Whitfield and Engineer found two distinct types of temperature-dependence: while the temperature-dependent levels $\mathcal{E}_1(i_1)$ in the partition function (15) determine the probability that state i_1 is occupied, the $\mathcal{E}_1(i_1) - T\, d_T \mathcal{E}_1(i_1)$ averaged in (4) determine the contribution actually made to the total energy by that state when it is occupied [33]. Incidentally, Landsberg's method of temperature-dependent energy levels is briefly mentioned in Pathria and Beale's landmark tektbook on statistical mechanics (see ch. 3, footnote 1 in [34]).

In recent years, and independently of Landsberg's work, a *statistical mechanics and thermodynamics at strong coupling* has been developed on the basis of a *Hamiltonian of mean force* [30,35–37]. The Hamiltonian of mean force is indeed a special case of (1), where the Hamiltonian of $E_1(i_1, i_2)$ may be decomposed as $H_1(i_1, i_2) = H_1(i_1) + \mathcal{I}(i_1, i_2)$, such that the total Hamiltonian $H(i_1, i_2)$ is expressed as a sum of three terms:

$$H(i_1, i_2) = H_1(i_1) + \mathcal{I}(i_1, i_2) + H_2(i_2). \tag{6}$$

The terms H_1 and H_2 correspond, respectively, to the bare system and the bare environment, and $\mathcal{I}$ is the ineraction Hamiltonian between the system and the enviroment. Averaging H_1 and $\mathcal{I}$ over the bare environment H_2 results in a temperature-dependent Hamiltonian $\mathcal{H}_1$ of mean force for the system given by [30]:

$$\mathcal{H}_1(i_1) = -k_B T \ln \left[\frac{\sum\limits_{i_2} e^{-\{H_1(i_1)+\mathcal{I}(i_1,i_2)\}/k_B T}\, e^{-H_2(i_2)/k_B T}}{\sum\limits_{i_2} e^{-H_2(i_2)/k_B T}} \right]. \tag{7}$$

In a regime where the interactions $\mathcal{I}$ with the environment are negligible compared to the bare system's energy H_1, the Hamiltonian of mean force $\mathcal{H}_1$ reduces to the bare (and temperature-independent) H_1. In the following section, we show that the thermostatistical theory *at strong coupling* developed on the basis of this Hamiltonian of mean force [30] reproduces the subdivision potential proposed by Hill in his thermodynamics of small systems [20].

3. Strong Coupling and Hill's Nanothermodynamics

In this section, we show that statistical mechanics at strong coupling provides a thermostatistical basis to Hill's nanothermodynamics. We start by providing a modest introduction to Hill's thermodynamic theory. We then use statistical mechanics at strong coupling to show that the subdivision potential in Hill's theory is reproduced when one considers the temperature-dependence of the effective Hamiltonian which results from the interaction between the system and the environment.

3.1. Hill's Subdivision Potential

In his nanothermodynamics [20], Hill starts by considering a traditional, homogeneous thermodynamic system with temperature T, pressure p and chemical potential μ. As this system is abstractly divided into smaller and smaller identical subsystems, the internal energy of each subsystem becomes eventually comparable to the energy of interaction amongst the subsystems, and, in contrast to traditional additive thermodynamics, the interaction can no longer be neglected. A new pair of conjugate variables emerges, namely the amount $\mathcal{M}$ of identical subsystems, and the *subdivision potential $\mathcal{E}$*. This subdivision potential may be thought of as the difference between a subsystem's true internal energy and the energy it would have if the rest of the subsystems were absent. For sufficiently large subsystems, the subdivision potential $\mathcal{E}$ is negligible.

After the subdivision process, each of the $\mathcal{M}$ identical subsystems has internal energy $\mathcal{U}$, entropy $\mathcal{S}$, volume $\mathcal{V}$ and $\mathcal{N}$ particles. Accounting for each subsystem's subdivision potential $\mathcal{E}$, the total energy of the system is given by a modified Euler equation:

$$\mathcal{MU} = T\mathcal{MS} - p\mathcal{MV} + \mu\mathcal{MN} + \mathcal{ME}. \tag{8}$$

And, for each individual subsystem, the energy becomes

$$\mathcal{U} = U + \mathcal{E}, \tag{9}$$

with U given by the usual Euler expression

$$U = TS - pV + \mu N. \tag{10}$$

Due to the incidence of the last term in (9), the internal energy $\mathcal{U}$ ceases to be a linear homogeneous function of $\mathcal{S}$, $\mathcal{V}$ and $\mathcal{N}$. This extra term, the subdivision potential $\mathcal{E}$, is the cornerstone of Hill's nonadditive nanothermodynamics. Paraphrasing Hill: *small-system thermodynamics departs from macroscopic thermodynamics in that [$\mathcal{U}$] is not a linear homogeneous function of [$\mathcal{S}$, $\mathcal{V}$ and $\mathcal{N}$]. Hence and extra term occurs in* (9). *This last two sentences epitomize the whole book* (see p. 24 in [20]).

When the subsystem of interest is an open system with a definite chemical potential μ and temperature T (imposed by the heat and particle reservoir made up of all other subsystems), then the additional energy term $\mathcal{E}$ in (9) must stem from an alteration in pressure. Interactions with the environment cause the system's pressure to deviate from p and become instead an effective pressure $\hat{p}$. Then, the system's internal energy is changed by an amount (see pp. 10, 24 in [20])

$$\mathcal{E} = -\left(\hat{p} - p\right)V. \tag{11}$$

If the subsystem of interest is, instead, a closed system with a definite pressure p and temperature T, then the additional energy must result from an alteration in chemical potential. Interactions with the environment cause the system's chemical potential to deviate from μ and become instead an effective $\hat{\mu}$. Then, the system's internal energy is altered by an amount (see pp. 16, 24 in [20])

$$\mathscr{E} = (\hat{\mu} - \mu)\,\mathcal{N}. \tag{12}$$

If the subsystem's only environmental constraint is the temperature T, then interactions with the environment will cause deviations in pressure (which becomes $\hat{p}$ instead of p) and chemical potential (which becomes $\hat{\mu}$ instead of μ). As a result, the additional subdivision energy $\mathscr{E}$ becomes

$$\mathscr{E} = -\,(\hat{p} - p)\,\mathcal{V} + (\hat{\mu} - \mu)\,\mathcal{N}, \tag{13}$$

or, as Hill wrote it (p. 24 in [20]),

$$\mathscr{E} = \mathcal{A} + p\mathcal{V} - \mu\mathcal{N}, \tag{14}$$

with $\mathcal{A} \equiv \mathcal{U} - T\mathcal{S} = \hat{\mu}\mathcal{N} - \hat{p}\mathcal{V}$. As the subsystem's size approaches the macroscopic limit, the difference between $(\hat{p}, \hat{\mu})$ and (p, μ) vanishes, and, as a result $\mathcal{U} \approx U \gg \mathscr{E}$ in every case.

3.2. Statistical Mechanics at Strong Coupling

Like in the previous section, we consider a small system with internal energy $\mathcal{U}$, volume $\mathcal{V}$ and $\mathcal{N}$ particles. For simplicity, we constrain the system's temperature to be T, while we allow the pressure and chemical potential to fluctuate as a result of interactions with the environment. The partition function $\mathcal{Z}_1$ may be written using the Hamiltonian of mean force (7):

$$\mathcal{Z}_1 = \sum_i e^{-\mathcal{H}_1(i)/k_B T}, \tag{15}$$

where, due to the nonvanishing interaction term in (7), $\mathcal{H}_1(i)$ is temperature-dependent. In the following, and for notational simplicity, we drop the subindex 1 used to label the system in Equation (7). The partition function (15) may be used to find the internal energy (4) of the system, resulting in a expression of the form (9), with

$$U \equiv \langle \mathcal{H}(i) \rangle, \tag{16}$$

and

$$\mathscr{E} \equiv -T \left\langle \frac{\partial \mathcal{H}(i)}{\partial T} \right\rangle. \tag{17}$$

The quantity U is the reference energy for the bare system in the absence of coupling. The additional term $\mathscr{E}$ is an excess energy resulting from strong interactions with the environment, which cause the effective Hamiltonian to be temperature-dependent. When the interaction $\mathcal{I}$ is absent (isolated system) or negligible (large system), Equation (7) simplifies and the derivative in (17) becomes zero.

The system's true pressure $\hat{p}$ is given by the usual

$$\hat{p} = -\frac{\partial \mathcal{U}}{\partial \mathcal{V}}, \tag{18}$$

or

$$\hat{p} = p + \Delta p, \tag{19}$$

where $p \equiv -\partial U/\partial \mathcal{V}$ is the pressure in the absence of coupling, and

$$\Delta p \equiv -\frac{\partial \mathscr{E}}{\partial \mathcal{V}} \tag{20}$$

is the additional pressure due to the interactions with the environment captured in $\mathscr{E}$. Likewise, the chemical potential $\hat{\mu}$ is given by

$$\hat{\mu} = \frac{\partial \mathcal{U}}{\partial \mathcal{N}}, \tag{21}$$

or

$$\hat{\mu} = \mu + \Delta\mu, \tag{22}$$

where $\mu \equiv \partial U / \partial \mathcal{N}$ is the chemical potential of the bare system, and

$$\Delta\mu \equiv \frac{\partial \mathscr{E}}{\partial \mathcal{N}} \tag{23}$$

is the additional chemical potential resulting from interactions. Combining (20) and (23), we see that the additional pressure and chemical potential change the internal energy of the strongly coupled system by an amount

$$\mathscr{E} = -\mathcal{V}\Delta p + \mathcal{N}\Delta\mu. \tag{24}$$

Expression (24) is equal to the subdivision potential (13) introduced by Hill for the canonical case using thermodynamic arguments. However, in contrast to Hill's purely thermodynamic approach, the subdivision potential has now a thermostatistical interpretation in terms of the temperature-dependence of the effective Hamiltonian (17). The method of statistical mechanics at strong coupling has recently been used to analyze interfacial phenomena, easily producing laws shown to be valid at the nanoscale [3].

Generalization to Other Ensembles

While the strong coupling analysis above was done for a system subject to canonical constraints, a similar statistical mechanical analysis may be carried out for other environmental variables.

In the grand canonical ensemble, the system has a fixed chemical potential, and the energy adjustment due to strong coupling is simply given by the first term in the right hand side of (24), which is the same as expression (11) of Hill's thermodynamic theory. Likewise, in the isothermal-isobaric ensemble, the pressure of the system is fixed, and the energy correction $\mathscr{E}$ due to strong coupling is given by the second term in the rhs of (24) alone, which is the same as expression (12) in Hill's theory.

The strong coupling contribution in each of the three ensembles is different, making them nonequivalent. However, for large systems, the coupling term $\mathcal{I}$ in (7) becomes negligible with respect to the bare system's Hamiltonian. As a result, the temperature dependence of the effective Hamiltonian vanishes, and, as expected, all ensemble descriptions become equivalent.

It should be noted that, in addition to the energy adjustment (17) needed to match the environment's temperature T, in the grand-canonical ensemble, the open system must also exchange particles in order to match the surrounding chemical potential μ. Therefore, the Boltzmann factor in (15) must be corrected with $e^{\mu n(i)}$, where $n(i)$ is the effective number of particles in the open system. A statistical mechanical analysis of this strongly coupled system results, not only in a modified energy $\mathcal{U} = U + \mathscr{E}$, but also a modified number of particles $\mathcal{N}$ consisting of the bare $\langle n(i) \rangle$, and an additional adjustment term $-T \langle \partial n(i) / \partial T \rangle$ resulting from the strong coupling.

Likewise, in the isothermal-isobaric ensemble, the small system must not only match the environmental temperature T by modifying its energy. It must also match the environmental pressure p by exchanging volume with its surroundings. Therefore, the Boltzmann factor in (15) must be augmented with $e^{-pV(i)}$, where $V(i)$ are effective volume states. This results, not only in a modified energy $\mathcal{U} = U + \mathscr{E}$, but also a modified volume $\mathcal{V}$ consisting of two terms: the bare system's volume $\langle V(i) \rangle$, and an additional term $-T \langle \partial V(i) / \partial T \rangle$ representing adjustments caused by strong coupling.

4. Concluding Remarks

The additivity of extensive quantities is an assumption upon which classical thermodynamics is based. However, this property does not apply to small systems [20] or, generally, to systems with sufficiently long-range interactions [1]. In this article, we have reviewed and shown the connection between three different theories proposed for the thermodynamic treatment of systems not satisfying the additivity ansatz. The differences and similarities between the three theories are illustrated in Figure 1.

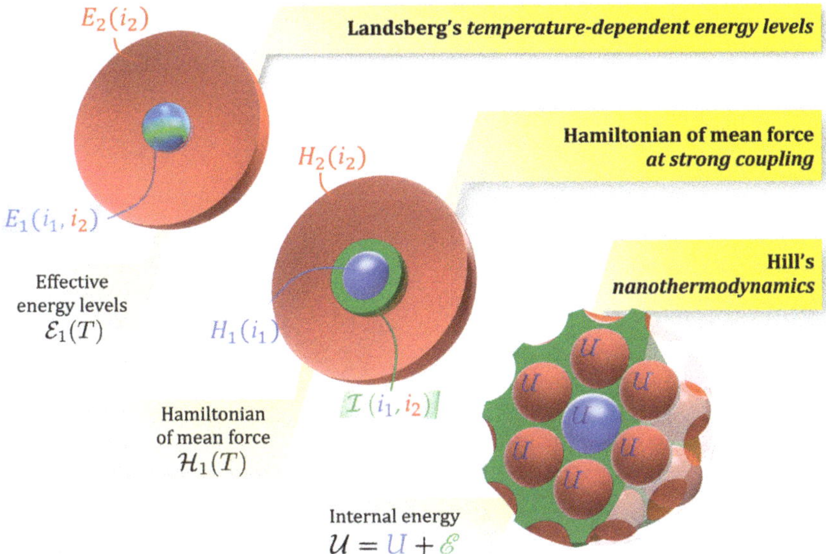

Figure 1. Illustration of the main features of the three small-system theories. In Landsberg's theory, the energy levels E_1 of the system (blue sphere) are modified into $\mathcal{E}_1$ due to its interaction with the (red) environment. Statistical mechanics at strong coupling starts with a total Hamiltonian consisting of the environment (red), the system (blue) and the interaction (green); averaging the system and the interaction over the environment results in a temperature-dependent Hamiltonian of mean force $\mathcal{H}_1$. Hill's nanothermodynamics considers the system of interest as if it were surrounded by a macroscopic set of interacting replicas; the interaction causes the effective energy to depart from U and become $\mathcal{U}$.

Landsberg's pioneering work consisted of assuming that when an otherwise isolated system comes into contact with a thermal bath, its energy levels are redistributed due to the interaction [6]. This results in an effective energy landscape that depends on the temperature of the bath, and it produces an extended thermostatistical framework that accounts for unknown interactions with the thermal environment (see Equations (4) and (5) above).

Years later, Hill proposed a thermodynamic framework for nonadditive systems by describing the mismatch between the internal energy of a small system when it is isolated vs. when it is embedded in a macroscopic ensemble of identical replicas. To account for this difference, he introduced the concept of subdivision potential, a new thermodynamic variable also known as *replica energy*. See [20,23] and Section 3.1 above.

More recently, a thermostatistical framework *at strong coupling* [30] has been formulated to account for the presence of the environment through a Hamiltonian of mean force (7). We show this modified Hamiltonian yields a temperature-dependent energy landscape as earlier suggested by Landsberg (see Equation (2)), and that it provides a thermostatistical foundation (17) for Hill's subdivision potential $\mathscr{E}$ as it makes its appearance in the internal energy (9). Furthermore, we show that the thermostatistical treatment of the temperature-dependent Hamiltonian of mean force in the canonical ensemble results in (24), which is the subdivision potential (13) obtained by Hill. Hence, we assert that statistical mechanics at strong coupling is the bridge between Landsberg's 'temperature-dependent energy levels' and Hill's nanothermodynamics. Indeed, and despite their independent inception and evolution, these three powerful theories describe similar phenomena, namely the nonadditivity caused by strong interactions.

Author Contributions: Conceptualization and Formal Analysis, R.d.M.; Supervision and Validation, J.M.R., R.d.M.; Writing, review & editing, R.d.M. Both authors have read and agreed to the published version of the manuscript.

Funding: J.M.R. is grateful to the Research Council of Norway for its Center of Excellence funding scheme, project No. 262644, PoreLab.

Conflicts of Interest: The authors declare no conflict of interest.

References

1. Campa, A.; Dauxois, T.; Fanelli, D.; Ruffo, S. *Physics of Long-Range Interacting Systems*; Oxford University Press: Oxford, UK, 2014.
2. Latella, I.; Peréz-Madrid, A.; Campa, A.; Casetti, L.; Ruffo, S. Thermodynamics of Nonadditive Systems. *Phys. Rev. Lett.* **2015**, *114*, 230601. [CrossRef] [PubMed]
3. de Miguel, R.; Rubí, J.M. Strong Coupling and Nonextensive Thermodynamics. *Entropy* **2020**, *22*, 975. [CrossRef] [PubMed]
4. Rushbrooke, G.S. On the statistical mechanics of assemblies wose energy-levels depend on temperature. *Trans. Faraday Soc.* **1940**, *36*, 1055. [CrossRef]
5. Landsberg, P.T. Statitical Mechanics of Teperature-Dependent Energy Levels. *Phys. Rev.* **1954**, *95*, 643.
6. Elcock, E.W.; Landsberg, P.T. Temperature Dependent Energy Levels in Statistical Mechanics. *Proc. Phys. Soc. Lond. Sect. B* **1957**, *70*, 161. [CrossRef]
7. Cuden, C.B. The Temperature Dependence of the Energy Gaps in Semiconductors. Ph.D. Thesis, University of British Columbia, Vancouver, BC, Canada, 1969. [CrossRef]
8. Bendt, P.J.; Cowan, R.D.; Yarnell, J.L. Excitations in Liquid Helium: Thermodynamic Calculations. *Phys. Rev.* **1959**, *113*, 1386. [CrossRef]
9. Donnelly, R.J.; Roberts, P.H. A theory of temperature-dependent energy levels: Thermodynamic properties og He II. *Low Temp. Phys.* **1977**, *27*, 687. [CrossRef]
10. Allen, P.V.; Heine, V. Theory of the temperature dependence of electronic band structures. *J. Phys. C Solid State Phys.* **1976**, *9*, 2305. [CrossRef]
11. Patrick, C.E.; Giustino, F. Unified theory of electron-phonon renormalization and phono-assisted optical absorption. *J. Phys. Condens. Matter* **2014**, *26*, 365503. [CrossRef]
12. Dykman, M.I.; Kono, K.; Kostantinov, D.; Lea, M.J. Ripplonic Lamb Shift for Electrons on Liquid Helium. *Phys. Rev. Lett.* **2017**, *119*, 256802. [CrossRef]
13. Erez, A.; Meir, Y. Effect of amplitude fluctuations on the Berezinskii-Kosterlitz-Thouless transition. *Phys. Rev. B* **2013**, *88*, 184510. [CrossRef]
14. Kolář, M.; Ryabov, A.; Filip, R. Optomechanical oscillator controlled by variatioons in its heat bath temperature. *Phys. Rev. A* **2017**, *95*, 042105. [CrossRef]
15. Yamano, T. Efficiencies of thermodynamics when temperature-dependent energy levels exist. *Phys. Chem. Chem. Phys.* **2016**, *18*, 7011. [CrossRef]
16. Yamano, T. Effect of temperature-dependent energy levels on exergy. *J. Phys. Commun.* **2017**, *1*, 055007.
17. de Miguel, R.; Rubí, J.M. Thermodynamics far from the thermodynamic limit. *J. Phys. Chem. B* **2017**, *121*, 10429–10434.
18. Shental, O.; Kanter, I. Shannon meets Carnot: Generalized second thermodynamic law. *Europhys. Lett.* **2009**, *85*, 10006. [CrossRef]
19. Hill, T.L. Thermodynamics of Small Systems. *J. Chem. Phys.* **1962**, *36*, 3182. [CrossRef]
20. Hill, T.L. *Thermodynamics of Small Systems, Parts I & II*; Dover: New York, NY, USA, 2013.

21. Hill, T.L. Perspective: Nanothermodynamics. *Nano Lett.* **2001**, *1*, 111. [CrossRef]
22. Hill, T.L. A different Approach to Nanothermodynamics. *Nano Lett.* **2001**, *1*, 273. [CrossRef]
23. Bedeaux, D.; Kjelstrup, S.; Schnell, S.K. *Nanothermodynamics—General Theory*; PoreLab: Trondheim, Norway, 2020.
24. Rubí, J.M.; Bedeaux, D.; Kjelstrup, S. Thermodynamics for Single-Molecule Stretching Experiments. *J. Phys. Chem. B* **2006**, *110*, 12733. [CrossRef]
25. Quian, H. Hill's small systems nanothermodynamics: A simple macromolecular partition problem with a statistical perspective. *J. Biol. Phys.* **2012**, *38*, 201. [CrossRef]
26. Chamberlin, R.V. The BigWorld of Nanothermodynamics. *Entropy* **2015**, *17*, 52. [CrossRef]
27. Galteland, O.; Bedeaux, D.; Hafskjold, B.; Kjelstrup, S. Pressures Inside a Nano-Porous Medium. The Case of a Single Phase Fluid. *Front. Phys.* **2019**, *7*, 60. [CrossRef]
28. Rauter, M.T.; Galteland, O.; Erdõs, M.; Moultos, O.A.; Vlugt, T.J.H.; Schnell, S.; Bedeaux, D.; Kjelstrup, S. Two-Phase Equilibrium Conditions in Nanopores. *Nanomaterials* **2020**, *10*, 608. [CrossRef] [PubMed]
29. Strøm, B.A.; He, J.; Bedeaux, D.; Kjelstrup, S. When Thermodynamic Properties of Adsorbed Films Depend on Size: Fundamental Theory and Case Study. *Nanomaterials* **2020**, *10*, 1691. [CrossRef]
30. Talkner, P.; Hanggi, P. Colloquium: Statistical mechanics and thermodynamics at strong coupling: Quantum and classical. *Rev. Mod. Phys.* **2020**, *92*, 041002. [CrossRef]
31. Kolář, M.; Ryabov, A.; Filip, R. Heat capacities of thermally manipulated mechanical oscillator at strong coupling. *Sci. Rep.* **2019**, *9*, 10855. [CrossRef] [PubMed]
32. de Miguel, R.; Rubí, J.M. Negative thermophoretic force in the strong coupling regime. *Phys. Rev. Lett.* **2019**, *123*, 200602. [CrossRef] [PubMed]
33. Whitfield, G.; Engineer, M. Temperature dependence of the polaron. *Phys. Rev. B* **1975**, *12*, 5472. [CrossRef]
34. Pathria, R.K.; Beale, P.D. *Statistical Mechanics*, 3rd ed.; Elsevier: Oxford, UK, 2011.
35. Talkner, P.; Hanggi, P. Open system trajectories specify fluctuating work but not heat. *Phys. Rev. E* **2016**, *94*, 022143. [CrossRef]
36. Seifert, U. First and Second Law of Thermodynamics at Strong Coupling. *Phys. Rev. Lett.* **2016**, *116*, 020601. [CrossRef] [PubMed]
37. Jarzynski, C. Stochastic and Macroscopic Thermodynamics of Strongly Coupled Systems. *Phys. Rev. X* **2017**, *7*, 011008. [CrossRef]

Publisher's Note: MDPI stays neutral with regard to jurisdictional claims in published maps and institutional affiliations.

MDPI

St. Alban-Anlage 66

4052 Basel

Switzerland

Tel. +41 61 683 77 34

Fax +41 61 302 89 18

www.mdpi.com

Nanomaterials Editorial Office

E-mail: nanomaterials@mdpi.com

www.mdpi.com/journal/nanomaterials